Springer Theses

Recognizing Outstanding Ph.D. Research

Aims and Scope

The series "Springer Theses" brings together a selection of the very best Ph.D. theses from around the world and across the physical sciences. Nominated and endorsed by two recognized specialists, each published volume has been selected for its scientific excellence and the high impact of its contents for the pertinent field of research. For greater accessibility to non-specialists, the published versions include an extended introduction, as well as a foreword by the student's supervisor explaining the special relevance of the work for the field. As a whole, the series will provide a valuable resource both for newcomers to the research fields described, and for other scientists seeking detailed background information on special questions. Finally, it provides an accredited documentation of the valuable contributions made by today's younger generation of scientists.

Theses may be nominated for publication in this series by heads of department at internationally leading universities or institutes and should fulfill all of the following criteria

- They must be written in good English.
- The topic should fall within the confines of Chemistry, Physics, Earth Sciences, Engineering and related interdisciplinary fields such as Materials, Nanoscience, Chemical Engineering, Complex Systems and Biophysics.
- The work reported in the thesis must represent a significant scientific advance.
- If the thesis includes previously published material, permission to reproduce this must be gained from the respective copyright holder (a maximum 30% of the thesis should be a verbatim reproduction from the author's previous publications).
- They must have been examined and passed during the 12 months prior to nomination.
- Each thesis should include a foreword by the supervisor outlining the significance of its content.
- The theses should have a clearly defined structure including an introduction accessible to new PhD students and scientists not expert in the relevant field.

Indexed by zbMATH.

More information about this series at https://link.springer.com/bookseries/8790

Wenrui Zhang

Oxide-Ion Conduction in the Dion–Jacobson-Type Layered Perovskites

Doctoral Thesis accepted by
Tokyo Institute of Technology, Tokyo, Japan

Springer

Author
Dr. Wenrui Zhang
Tokyo Institute of Technology
Tokyo, Japan

Supervisor
Prof. Masatomo Yashima
Tokyo Institute of Technology
Tokyo, Japan

ISSN 2190-5053 ISSN 2190-5061 (electronic)
Springer Theses
ISBN 978-981-19-2246-6 ISBN 978-981-19-2247-3 (eBook)
https://doi.org/10.1007/978-981-19-2247-3

This Springer imprint is published by the registered company Springer Nature Singapore Pte Ltd.
The registered company address is: 152 Beach Road, #21-01/04 Gateway East, Singapore 189721, Singapore

Supervisor's Foreword

Zhang's Ph.D. research focused on the discovery of new structural family of oxide-ion conductors through theoretical and experimental methods. The characterization of the electrical and structural properties of the new structural family of oxide-ion conductors, which belongs to the Dion–Jacobson-type layered perovskite family, has been investigated. Such materials are hard to come by, so the discovery of these materials, the new method and the proposed concept will make the realization of many environment-friendly technologies.

Firstly, Dr. Zhang's doctoral thesis described the theoretical method which is the bond valence-based (BV) energy calculation. He has explored Dion–Jacobson-type oxide-ion conductors by screening the substances with low energy barriers (E_b) for oxide-ion migration. The E_b is calculated by the BV method, which is a quick and efficient technique for screening potential oxide-ion conductors, compared with other techniques such as molecular dynamics simulations and density functional theory calculations due to its simple principle. The $CsBi_2Ti_2NbO_{10}$ and $CsLn_2Ti_2NbO_{10}$ (Ln = La, Pr, Sm, and Nd) compositions are selected because the E_bs are relatively low.

Secondly, the thesis described the synthesis, characterization of structural and transport properties of the Dion–Jacobson phases. Experimental results have demonstrated that the Dion–Jacobson-type layered perovskite $CsBi_2Ti_2NbO_{10-\delta}$ is an excellent oxide-ion conductor. This is the first report of a fast oxide-ion conductor in Dion–Jacobson-type layered perovskite family. Surprisingly, it is found that the bulk conductivities of Dion–Jacobson phase $CsBi_2Ti_2NbO_{10-\delta}$ are higher than those of the conventional yttria-stabilized zirconia (YSZ) and comparable to those of the best oxide-ion conductors. Lots of techniques have been carried out to verify the oxide-ion conduction in the Dion–Jacobson phase $CsBi_2Ti_2NbO_{10-\delta}$. In addition, $CsLa_2Ti_2NbO_{10-\delta}$ is the first example of a Dion–Jacobson-type oxide-ion conductor without phase transitions. No phase transition indicates good practical applications and strengthens the potential of Dion–Jacobson phases as practical oxide-ion conductors. The conductivities of Dion–Jacobson phase $CsBi_2Ti_2NbO_{10-\delta}$ and $CsLa_2Ti_2NbO_{10-\delta}$ have also been improved by the substitution successfully.

At last, the structural origin of the high oxide-ion conductivity in Dion–Jacobson phases has been presented. The oxygen ion transport is attributable to the large anisotropic thermal motions of oxygen atoms, the presence of oxygen vacancies, and the formation of oxide-ion conducting layers in the crystal structure. In addition, a new concept is also proposed; large bottlenecks for oxide-ion migration by large size of Cs^+ and Bi^{3+} (La^{3+}) displacement. The Cs^+ expands the Bi−Bi (La−La) distance along the *b*-axis due to the cation ordering of larger $Cs^+$and smaller Bi^{3+} La^{3+}). The Bi^{3+} (La^{3+}) cation is displaced along the *c* axis apart from the Ti/Nb1−O1 layer, which can be explained by the electrostatic forces. The Bi^{3+} (La^{3+}) displacement expands the Bi−Bi (La−La) distance along the *c* axis. The expansions of Bi−Bi (La−La) distances along both the b ($= a$) and c axes lead to large bottleneck sizes for oxide-ion diffusion.

Overall, this thesis is valuable and the present finding in this work of high oxide-ion conductivities in the new structure family, Dion−Jacobson phases, and the proposed concept would facilitate the design of novel oxide-ion conductors based on the Dion−Jacobson phases. Furthermore, the findings of this study open up the possibilities for many novel applications that will lead to a sustainable future.

Tokyo, Japan
February 2022

Masatomo Yashima

Parts of this thesis have been published in the following journal articles:

1. Wenrui Zhang, Kotaro Fujii, Eiki Niwa, Masato Hagihala, Takashi Kamiyama, Masatomo Yashima. Oxide-ion conduction in the Dion–Jacobson phase $CsBi_2Ti_2NbO_{10-\delta}$. *Nature communications*. 2020, *11*, 1224.
2. Wenrui Zhang, Kotaro Fujii, Tomohiro Ishiyama, Harue Kandabashi, Masatomo Yashima, Dion–Jacobson-type oxide-ion conductor $CsLa_2Ti_2NbO_{10-\delta}$ without phase transitions. *Journal of Materials Chemistry A*. 2020, 8.47: 25085-25093.
3. Wenrui Zhang, Masatomo Yashima, Improved oxide-ion conductivity by substitution of Sr for Bi in Dion-Jacobson phase $CsBi_2Ti_2NbO_{10}$, *Ceramics International*. 2022, in press. doi.org/10.1016/j.ceramint.2022.02.059.

Acknowledgements

I cordially wish to express my gratitude to my supervisor Prof. M. Yashima for his invaluable advice and useful suggestions over the whole works in the study. I truly appreciated the fact that his office door was always open for me from morning to night even at weekends. I have learned so much from him, such as how to make scientific writing logically and how to make a clear presentation. I have received many helps from him to go through the difficulty encountered when living abroad. He also given me a lot of kind helps and suggestions for the job-hunting. I will always be grateful for the positive influence and considerate helps from him.

I would like to thank Prof. M. Kawano, Prof. K. Maeda, Prof. Y. Okimoto, and Prof. H. Uekusa of Tokyo Institute of Technology for their useful comments and fruitful discussions on the thesis paper. I wish to thank Dr. K. Fujii and Dr. E. Niwa of Tokyo institute of Technology and all members of Prof. Yashima's laboratory for their kind-hearted assistance for sharing joyful memories during my work and study. I gratefully acknowledge the funding received from the "Tonen International Scholarship Foundation" and the "Japan Student Services Organization Scholarships". I am very grateful to my supervisor during my master's degree, Prof. Tingli Ma, Kyushu Institute of Technology for her invaluable advice and useful suggestions.

I am very grateful to Dr. S. Torii, Dr. P. Miao, and Dr. S. Kawaguchi for the assistance in the neutron and synchrotron X-ray diffraction experiments. I express special thanks to Mr. M. Shiraiwa, Miss H. Kandabashi, and Mr. K. Hibino for useful discussion and assistance in the experiments. I acknowledge Daiichi Kigenso Kagaku Kogyo Co., Ltd. for the AAS and ICP-OES measurements. The neutron-diffraction measurements were carried out by the project approval (J-PARC MLF Proposal No. 2017A0111 and 2017L1301). The synchrotron experiments were carried out at SPring-8 (Proposal No. 2017A1803, 2017B1265, 2018A1259, 2018B1056, and 2019A1052). This study was partly supported by Grants-in-Aid for Scientific Research (KAKENHI, No. JP15H02291, JP16H00884, JP16H06293, JP16H06440, JP16H06441, JP16H06438, JP16K21724, JP17K17717, JP17H06222, and JP19H00821) from the Ministry of Education, Culture, Sports, Science and Technology of Japan and JSPS Core-to-Core Program, A. Advanced Research Networks (Solid Oxide Interfaces for Faster Ion Transport).

Finally, I sincerely appreciate my wife Dr. Mu Li and my family for encouraging and supporting me in what I always wanted to do. Thanks for their indulgence, tolerance, and love all the time. I am particularly grateful for the contributions so much to me.

Contents

Chapter 1
Introduction

1.1 Background

1.1.1 Energy and the Environment

Global environmental issues, energy issues, and resources are the biggest challenges for our society. Durable, clean, and renewable energy devices are essential to global economic growth and human development, which presents a huge challenge. This challenge is made even more difficult due to the environmental threat of global warming, which is mostly caused by the carbon dioxide (CO_2) emissions produced from the primary energy sources, such as fossil fuels [1–3]. In order to stop the increase of CO_2, which is considered to be a cause, it is required to improve the coefficient of performance and sophisticating methods of energy supply structures [4, 5]. Energy production from wind power and photovoltaics could really help to avoid this dramatic scenario, but the fluctuating power from wind or sun could destabilize the future energy grids [6–9].

As a result, further green energy conversion and storage technologies are required. Smart communities that place small-scale distributed power generation systems near electricity demand sites and make high use of energy are drawing attention. Power generation systems based on renewable energy such as solar cells are considered the core of the smart community, however, it is difficult to supply stable electric power in many cases such as at nighttime or in unstable weather. Therefore, a distributed power generation system that utilizes fuel with high energy-conversion efficiency is expected to be an effective solution. Because the power generation efficiency using natural gas was achieved at 60%, the high-efficiency power generation system is introduced to the coastal area near the large-scale electricity. However, due to the far distance power transmission from coastal thermal power plants to the inland area, the power transmission losses cannot be ignored. Thus, a fuel cell is expected as a high-efficiency distributed power generation system [1, 3, 10–12].

W. Zhang, *Oxide-Ion Conduction in the Dion–Jacobson-Type Layered Perovskites*, Springer Theses, https://doi.org/10.1007/978-981-19-2247-3_1

Fuel cells should be one of these alternatives, thanks to their well-known high theoretical efficiency and low emission of pollutants. Their value as an effectual solution to more sustainable ways of energy production was already foreseen decades ago. Fuel cells are devices that convert energy from chemicals into electricity [13, 14]. The fuels (like H_2, CO) are oxidized at the anode part, thus producing electrons and then forced to the cathode part, where the oxidant species are reduced there. The charged species (ions), generated at either the anode or the cathode, move through an electrolyte to the opposite electrode. The passage of the electrons via the external circuit generates electric current [2, 11, 15].

Nowadays, there are many different fuel cell systems, which are classified by the used electrolyte type conveniently, such as polymeric-electrolyte membrane fuel cells (PEMFC), high-temperature solid oxide fuel cells (SOFCs), phosphoric acid fuel cells (PAFCs), molten carbonate fuel cells (MCFCs), and alkaline fuel cells (AFCs) [13, 14, 16]. SOFCs employ a solid inorganic oxide material as an electrolyte. The electrolyte is sandwiched between the anode and cathode electrodes, which are mainly used to transport the oxide ions produced from the cathode part to the anode part (Fig. 1.1). O^{2-}-ions are generated due to the oxygen reduction in the cathode part, and then travel through the electrolyte to the anode part. Here, the oxide ions react with the fuel (H_2 for simplicity) and generate electrons. The electrons then flow back to the cathode, and thus produce an electric current [11, 17].

SOFCs are particularly attractive due to their various features. First, it facilitates the design and manufacture of different components due to the solid-state devices, which can lift liquid electrolytes limitation. Likewise, the modularity of the latter permits the combination of multiple cells, thus generating "stacks" that can be arranged together to produce the desired power output. Consequently, SOFCs have a potential application ranging from portable apparatus, automotive sector to stationary power stations for residential, commercial, and industrial uses. Meanwhile, it is possible to directly use the direct hydrocarbon fuels due to that the SOFCs must operate at very high temperatures. The use of external and expensive reform

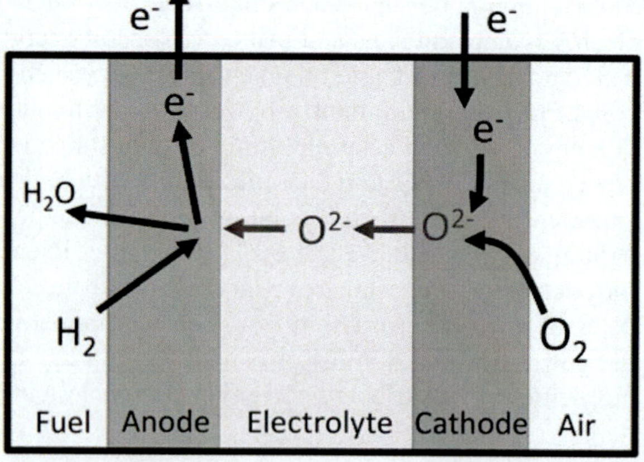

Fig. 1.1 Scheme of the solid-oxide fuel cells (SOFCs)

machinery can be avoided, and the range of fuels can be expanded. Furthermore, the heat produced as the secondary product can be used in cogeneration or combined with some heat systems, which are able to increase the total efficiency of the energy devices [17, 18].

However, despite the high capability of SOFCs, there are still some problems that should be resolved. Actually, the materials composing the different parts have to meet demanding requirements in terms of mechanical, thermal and redox stability, chemical and mechanical compatibility, and high electrical conductivity. In addition, due to the high temperature of typical SOFCs exercise [19], it accelerates the degradation of the components, slows down the start-up/shut-down cycles, reduces the theoretical efficiency, and increases the cost. Thus, it is necessary to decrease the operating temperature from high-temperature range to an intermediate (500–800 °C) or low range (<500 °C) [20–23]. However, operation at lower temperature causes other problems that increase electrolyte resistance and slow electrode kinetics. The problems could be solved with the careful selection and development of the electrodes and the electrolyte materials, where the latter are more important because they dictate the properties of the SOFC unit and affect the oxide ions' transport efficiency from the cathode to the anode [24, 25].

An SOFCs electrolyte should ideally show good stability at a high range of oxygen partial pressures, exhibits good thermal and mechanical stability, has negligible electronic conductivity and high ionic conductivity, and possess a fully dense structure to limit any reactant crossover. In addition, it is desirable to operate at lower temperatures. To date, numerous oxide-ion conductors have been discovered to overcome these requirements; but, all of them have some limitations. Thus, it is still essential to develop a better oxide-ion conductor. Because the oxide-ion conductivities are greatly affected by the structure. Therefore, the structural properties of the ion conductors should be concerned. The structural properties should be synthesized and controlled thoughtfully during the discovery and development of new ion conductors. To understand the oxide-ion conduction mechanisms, structural and electrical characterizations are performed in conjunction with computer modeling techniques, which is important to improve the electrochemical properties of the electrolyte for wider practical applications [11, 16, 18, 25].

1.1.2 Oxide-Ion Conductor

Oxide-ion conduction in solids was discovered by Nernst more than 100 years ago [27]. Currently, the applications of the oxide-ion conductors are being studied, such as oxygen separation membranes, oxygen sensors, catalytic supports, and electrolytes for SOFCs [25, 28–30]. As mentioned before, the characteristics of the electrolytic material greatly affect the properties of an SOFC. The ideal ionic conductor can efficiently transport oxide ions at lower temperatures. However, the ionic transport in these materials is strongly thermally activated due to the large size of the O^{2-} anion. In general, the conduction mechanism takes place by the mobile oxygen ions hopping through vacant or interstitial sites, which is assisted by the disorder of the structure. This suggests crystal structure's crucial role in the definition of properties of this kind of ionic conductor, which are grouped according to their structural features. Various oxide-ion-conducting materials will be presented afterward, with particular attention to their structural properties, which aim to present the main oxide ion-conducting materials and to highlight the common structural characteristics at the basis of the oxide-ion conduction. It is an important material expected to be applied to electrolytes, oxygen sensors and oxygen permeable membranes, and SOFCSs. In particular, there are some requirements for the electrolyte of SOFCs [9, 31–34].

1. Stability under a wide range of oxygen partial pressure and thermochemical stability at high temperature.
2. Excellent ionic conductivity.
3. High transportation (neglectable electron conductivity).

Because oxide-ion conduction is closely related to the crystal structure, it is necessary to study their crystal structure. YSZ (yttria-stabilized zirconia) is frequently utilized as an electrolyte of SOFCs. ZrO_2 belongs to the monoclinic at room temperature, which transitioned to distorted fluorite-type structure at about 1200 °C, and transition to cubic fluorite-type structure at high temperature. However, by replacing Zr^{4+} with Y^{3+}, the structure can be stable cubic fluorite even at room temperature and shows no change with the temperature increasing. Furthermore, the substitution of Y^{3+} increases the oxygen vacancy concentration, which is a carrier of oxide-ion conduction. YSZ shows high ionic conductivities at high temperature, good chemical and mechanical stability, and an ionic transport number higher than 0.99. However, there is a problem that the ionic conductivity of YSZ in the medium-temperature range is low. In order to solve this problem, a thin YSZ film has been developed as an anode-supported cell structure of medium-temperature SOFCs. Oxides with a fluorite crystal structure like YSZ, or bi-based fluorite crystal structures like scandia-stabilized zirconia and Bi_2O_3 have high ionic conductivities. Recently, the ionic conductors having various structures such as perovskite, perovskite related, apatite, and melilite type have been actively studied and developed [8, 16, 26, 35–37].

1.1.3 Perovskite and Perovskite-Related Oxides

1.1.3.1 Single Perovskites

Perovskite "$CaTiO_3$" was discovered by Rose and named for Russian Mineralogist Count Lev Aleksevich Perovski. The perovskite structure family has a similar structure related to the mineral perovskite of $CaTiO_3$. Ideal perovskite structures usually show the chemical formula of ABX_3, where the *A*-site can be filled with either alkali, alkali earth, or transition metal cations (coordination number: 12); *B*-site can be filled with transition metal cations (coordination number: 6); and *X*-site is anion like oxygen O, nitrogen N (Fig. 1.2). They show very wide chemical and physical

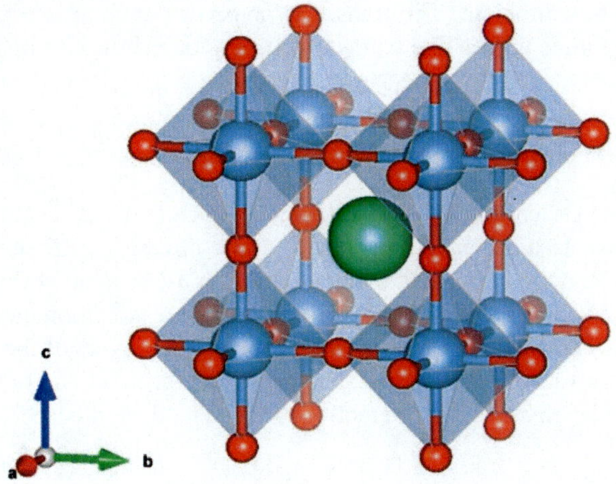

Fig. 1.2 The crystal structure of an ideal cubic perovskite ABO_3

properties like photocatalytic, magnetic, ionic conduction, super-conduction, ferroelectricity, etc. Nature of *A* and *B*, like their ionic sizes and, or charges, distortions, tilting, and position displacement may lead to new properties [38–40].

Almost both of the elements can be accommodated in the perovskite or perovskite-related structure, resulting in many derivatives, polymorphs, and super-structures. Thus, these materials are probably the widest studied systems in the world, with fascinating characteristics. Perovskite compounds are the main components of the core of our earth, and have colossal magneto resistivity, ferroelectricity, superconductivity, piezoelectricity, catalytic activity, and can be applied as photovoltaic materials. Furthermore, oxides of the perovskite family also show electronic as well as ionic conductivity. Currently, various fast oxide-ion conductors were reported possessing the perovskite or perovskite-related structures, and are being studied for their potential use as SOFC electrolytes [41].

For the ideal ABO_3 perovskite, *B* cations are in 6-coordinated sites forming a backbone of BO_6 octahedra, and interstitial voids are occupied by the large *A* cation in a 12-coordination with the O. The structure is cubic and crystallizes in the $Pm\overline{3}m$ space group. Tilting and or rotation of the octahedron cause distorted variants with lower symmetries. The tolerance factor rationalizes this phenomenon, which is given by

$$t = \frac{r_A + r_O}{\sqrt{2}(r_B + r_O)} \tag{1.1}$$

For an ABO_3 perovskite oxide, the r_A, r_B, and r_O denote ionic radii of *A* cation, *B* cation, and oxygen, respectively. If the tolerance factor was 1, the structure is a perfect perovskite cubic without any distortion; whereas if the tolerance factor was smaller or larger than 1, the octahedra will become tilting. Octahedral factor ($= r_B/r_O$) is another factor to show the formation of the perovskite structure. The perovskites should be in the range of octahedral factor with 0.414–0.732, which is necessary but not sufficient. The transition involves nonpolar lattice distortions, leading to polar lattice modes like trilinear coupling or hybrid improper mechanisms [41].

1.1.3.2 Double Perovskites

A structure consisting of layers of ABO_3 and or $A'B'O_3$ leads to a tetragonal symmetry, such a super-lattice as an *A*-site-ordered or *B*-site-ordered double perovskite. Rondinelli reported the *A*-site cations ordering of $AA'B_2O_6$ double perovskites. Double perovskites with the $AA'BB'O_6$ stoichiometry, such as $NaLaMgWO_6$ and $NaLaScNbO_6$, tend to show the ordering layers of the *A*-site and A'-site cations and or of the *B*-site and B'-site cations (Fig. 1.3). The ordering arrangements optimize the electrostatic interactions well. The design based on the ordering concept could

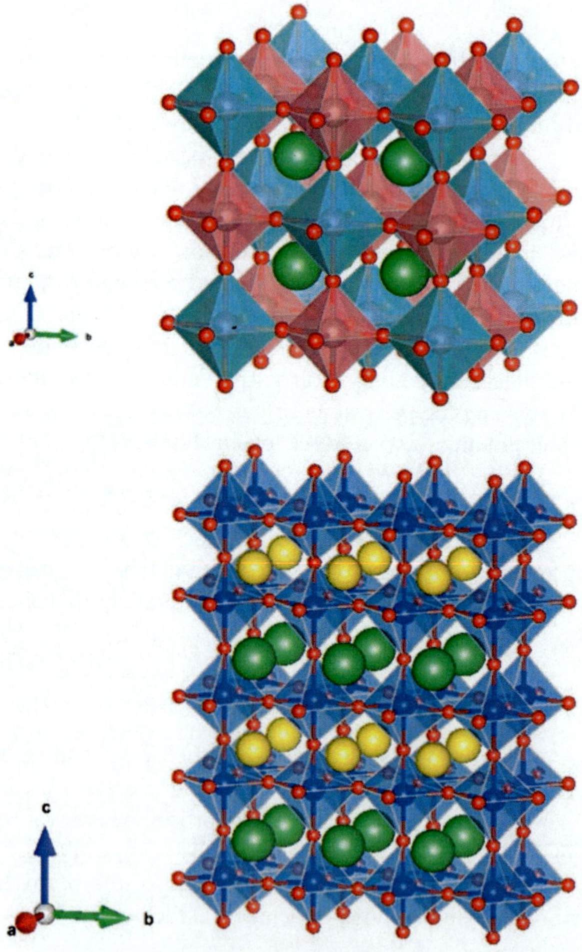

Fig. 1.3 The crystal structure of double-layered perovskites of $AA'B_2O_6$ (*A*-site ordering) and $A_2BB'O_6$ (*B*-site ordering), which were derived from the cubic ABO_3. Double perovskites tend to exhibit layered ordering of the *A* and A' cations, or layered ordering of rock salt ordering of the *B* and B'

be extended to other double perovskites. In the RLaNiMnO$_6$ (R are rare-earth ions) layered perovskite oxides, the polarization increases because of the difference in the ionic radii between the rare-earth ions and La cations, which leads to higher octahedral tilts.

1.1.3.3 Ruddlesden–Popper Phases

Ruddlesden–Popper phase, showing a chemical formula of $A_{n+1}B_nO_{3n+1}$, usually forms a homologous series, which is described as a stacking of perovskite-blocks layers, and with an extra rock-salt layer inserted in each perovskite-blocks layer (Fig. 1.4). Hence, the BO_6 octahedra in Ruddlesden–Popper phase are connected just in two dimensions. The lattice topology is directly responsible for a lot of chemical and or physical properties in the Ruddlesden–Popper phases, such as ferroelectricity and high ionic mobility. In the Ruddlesden–Popper phases, the A-site is metal cations usually with valences like 2 + and 3 + (such as Ba^{2+}, Ca^{2+}, La^{3+}) and the B-site is usually occupied by transition metal cations such as Ti^{4+}, Ni^{4+}, Nb^{5+}, and Mn^{2+}.

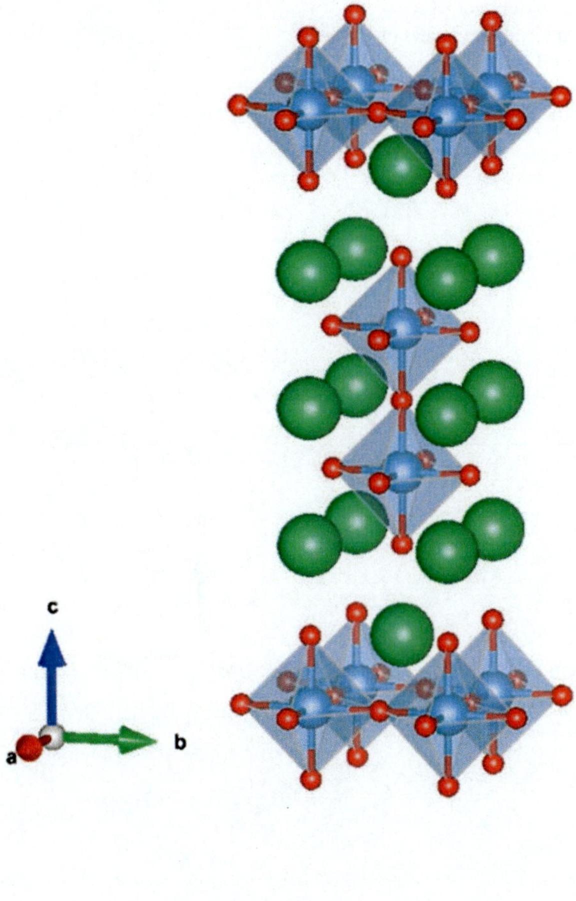

Fig. 1.4 The crystal structure of Ruddlesden–Popper-type layered perovskites of general formula $A_{n+1}B_nO_{3n+1}$, which were derived from the cubic ABO_3. This structure is usually described as a layered stacking of perovskite blocks with an extra rock salt AO layer inserted in every n perovskite unit cell along the [001] direction

1.1.3.4 Aurivillius Phases

Aurivillius phases, showing a chemical formula of $Bi_2A_{m-1}B_mO_{3m+3}$, also form a layered structure (Fig. 1.5). The structure consists of fluorite-like Bi_2O_2 layers, alternating with perovskite-blocks layers ($A_{m-1}B_mO_{3m+1}$). Typically, the A-site cations (12-coordinated) are cations such as alkali metals (Na^+, K^+), alkaline-earth metals (Ba^{2+}, Ca^{2+}), Bi^{3+} or rare earth metals, and the B-site cations (6-coordinated) are also the transition metal cations like Ti^{4+}, Nb^{5+}, W^{6+}, Mo^{6+}. Typical at high-temperature ranges, their crystal structures are the tetragonal $I4/mmm$ space group that could regard as the ideal high-symmetry crystal structure. However, at low-temperatures ranges, the symmetry may decrease and the majority of them adopt polar orthorhombic and or monoclinic structures.

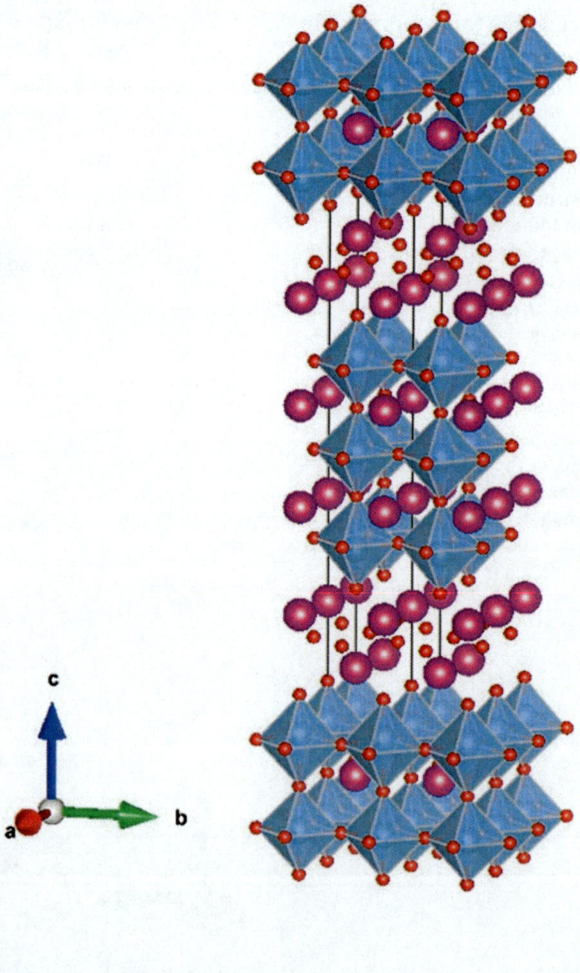

Fig. 1.5 The crystal structure of Aurivillius-type layered perovskites of $Bi_2A_{m-1}B_mO_{3m+3}$, which were derived from the cubic ABO_3. The Aurivillius structure consists of fluorite-like layers of Bi_2O_2 and perovskite-like $A_{m-1}B_mO_{3m+1}$ blocks along the [001] direction

1.1.3.5 Hexagonal Derivatives

In all the perovskite-based systems described before, the principal structural motif is the BO_6 octahedra connected by one or more of their vertices. However, it is possible to have perovskite structures in which the BO_6 octahedra are linked by their faces as shown in Fig. 1.6. In fact, when the cationic radius of A is particularly large (or the cationic radius of B particularly small), parts of the B-site cations share a face of the octahedron collectively with another octahedron. In this case of $t > 1$, the system possesses hexagonal symmetry. To describe these hexagonal derivatives, it is convenient to introduce a different way of visualizing the perovskite structure. The atomic arrangement could be defined using the close-packing hexagonal AO_3 layers (hexagonal symmetry; point group 6 mm) along a certain crystallographic direction (normally c), where the B cations are inserted into the AO_3 layers (see Fig. 1.6).

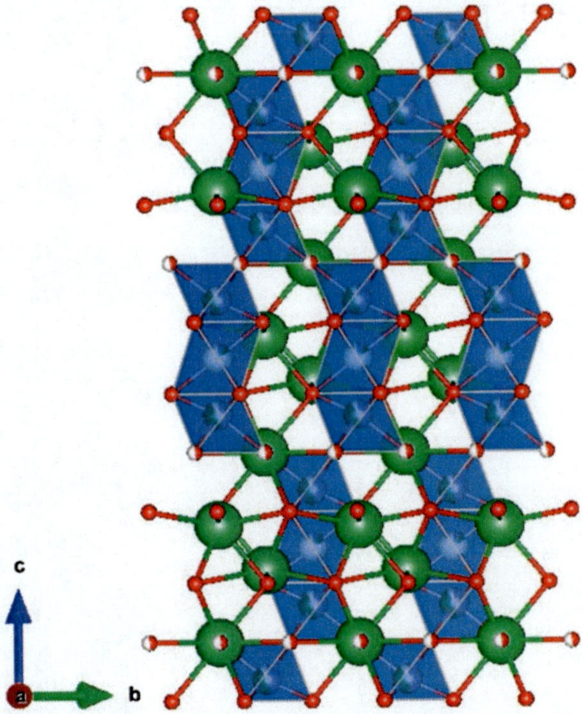

Fig. 1.6 The crystal structure of hexagonal perovskites. The structures are formed when the BO_6 octahedra share their faces. The different stacking sequences of hexagonal AO_3 layers and the cubic and hexagonal can lead to different unique crystal structures

1.1.3.6 Dion–Jacobson Phases

The Dion–Jacobson phases have the chemical formula of $A'[A_{n-1}B_nO_{3n+1}]$, where typically A' is an alkali metal like Li^+, Na^+, Cs^+; A is a rare-earth or alkaline-earth metal like La^{3+}, Ca^2, Bi^{3+}; B is a transition metal like Ti^{4+} Nb^{5+} Ta^{5+}, and n denotes the number of the BO_6 octahedral layers. The A' cations separate the $[A_{n-1}B_nO_{3n+1}]$ perovskite-like layers, leading to a layered perovskite structure (Fig. 1.7). The A' layers could also be a metal halide complex, such as NH^{4+}. Dion–Jacobson-type layered perovskites have attracted much attention because of their novel physical and chemical properties such as ferroelectricity, photocatalyst, magnetism, and lithium-ion conduction.

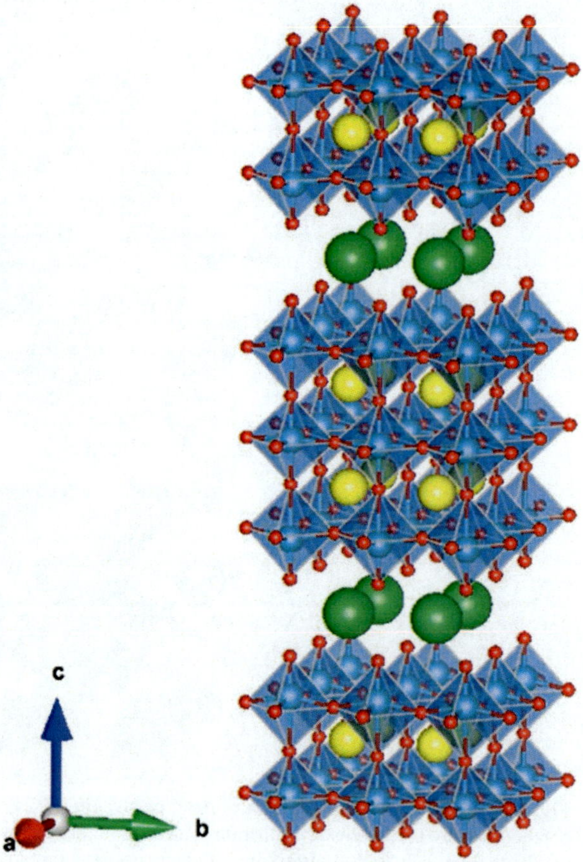

Fig. 1.7 The crystal structure of Dion–Jacobson-type layered perovskites of $A'[A_{n-1}B_nO_{3n+1}]$, which were derived from the cubic ABO_3. The structures homologous series with general formula $A'[A_{n-1}B_nO_{3n+1}]$ for $n > 2$

1.2 Previous Studies

1.2.1 Oxide-Ion Conduction in Perovskite or Perovskite-Related Structures

1.2.1.1 Perovskite Oxides of Doped Lanthanum-Gallate ($LaGaO_3$).

An example of a conductor possessing a perovskite structure is doped lanthanum-gallate ($LaGaO_3$). Doping using strontium and magnesium produces the $La_{1-x}Sr_xGa_{1-y}Mg_yO_{3-\delta}$ solid solution (usually referred to as LSGM), which shows excellent oxide-ion 4嗚4 + conductivities. The parent $LaGaO_3$ phase crystallizes with an orthorhombic phase and shows a phase transition from the orthorhombic to the rhombohedral system above 150 °C. Depending on the relative amount of the cationic dopants, the (Sr, Mg)-doped compounds can show monoclinic (space group *I2/a*), orthorhombic (*Pbnm*) or cubic symmetry even at room temperature. They then convert to higher symmetry phases, such as cubic or rhombohedral, with the temperature. The substitution of La^{3+} and Ga^{3+} with acceptor dopants with lower valences such as Sr^{2+} and Mg^{2+}, resulting in the oxygen vacancies according to the following reaction

$$2AO + 2M_M^x + O_O^x \rightarrow 2A'_M + V_O^{\cdot\cdot} + M_2O_3 \tag{1.2}$$

where A = Sr, Mg and M = La, Ga.

These vacancies are preferentially located in the first coordination shell of Ga, as evinced by an NMR study in which a certain amount of the Ga sites resonate with a shift consistent with the gallium in a fivefold coordination. Computational and experimental techniques have shown that the oxide-ion conduction is mediated using a hopping by the oxygen vacancies along the edges of GaO_6 octahedron, and subsequent relaxation of the neighboring cations. Various dopants have been tested for LSGM. Substituting La with other rare-earth cations linearly decreases the conductivity with the difference with the ideal ionic radii for an ideal cubic perovskite-type structure, i.e., using tolerance factor. A similar phenomenon happens by replacing Sr with K, Ca, or Ba. Whereas, partial substitution of Ga or Mg with other transition metals generally increases the electronic component of the conductivity. However, co-doping with small percentages of other cations like Fe and Co has been confirmed to benefit the performances of the substituted LSGM despite the slightly lower ionic conduction component [40–44].

Tests with various configurations and electrode materials show the potential of the LSGM systems as electrolytes for intermediate-temperature SOFCs. However, LSGM materials are difficult to synthesize, due to the secondary phases formation of $LaSrGa_3O_7$ and $LaSrGaO_4$ and the high sintering temperatures required for dense samples, which results in gallium oxide volatilization and undesired reactions with the anodic materials. Interlayers of SSZ, SDC, and La_2O_3-doped ceria are currently being used to stop the metal diffusion at the electrolyte–anode interface while alternative synthetic and sintering methods (such as precipitation and sol–gel synthesis, spark-plasma, and microwave sintering) are studied to obtain phase pure dense specimens at lower temperatures.

1.2.1.2 Perovskite Oxides of $Na_{0.5}Bi_{0.5}TiO_3$

Another famous single perovskite-type oxide-ion conductor is $Na_{0.5}Bi_{0.5}TiO_3$, which is also a ferroelectric material. Neutron diffraction results have shown that $Na_{0.5}Bi_{0.5}TiO_3$ exhibits a rhombohedral structure at room temperature, which then converts to a tetragonal structure around 320 °C, and to a cubic phase around 540 °C [45–47].

Detailed HRTEM studies suggest the complexity of the local structure, which is attributable to the ease of tilt of the TiO_6 octahedra. Domains of tilting *b* are in fact present in the whole temperature range and retain even in the high-temperature cubic structure. Instead of considering these domains as discretely interspersed within each other, a continuous tilting model is proposed. At a given temperature, a certain tilting model will be dominant, thus determining the average structural distortion, and hence the specific symmetry, depending on the diffraction techniques. The electrical properties of $Na_{0.5}Bi_{0.5}TiO_3$ are strongly affected by small variations in the *A*-site stoichiometry, which is also dependent on the starting Na and Bi amounts. When Bi/Na $\leq$ 1, bi-deficient compositions with high bulk conductivity are formed, also due to the unavoidable Bi_2O_3 loss during processing at high temperatures. On the contrary, if Bi/Na > 1, insulating ceramics (about 1×10^{-7} S cm^{-1} at the same temperature) are obtained. Bi-deficiency introduces oxygen vacancies that are considered the reason for the oxide ion migration. Computational studies evidence that these vacancies are preferentially formed in the vicinity of the Bi atoms (due to the weaker Bi–O bond), and the oxide ions most likely migrate from the mixed *A*-site layers to the oxygen vacancy-rich Bi layers via the edges of the TiO_6 octahedra. Overall, the higher polarizability of Bi cations together with the complexity of the lattice, is also an important factor in the oxide-ion conductivities.

The bulk conductivities could be enhanced using doping Mg on the titanium site. The grain boundary resistance above 500 °C and extends the oxide-ion conduction window to lower p_{O_2} values. However, computational work demonstrates that higher *B*-site doping could make an increase in the oxygen migration barriers, due to the binding of the vacancies with the substituents. It proposes that *A*-site doping is also an effective way to higher the oxide-in ionic conductivities, because the disordered Na/Bi lattice can form various local atomic configurations, and accommodate the electrostatic and strain fields due to the dopants.

Due to the high sensibility of the electrical properties to the initial composition, we need to do the preparation of NBT materials carefully. It shows that the water content of the raw materials affects the stoichiometry of the final products; for example, using moist Na_2CO_3 leads to insulating Bi-rich compositions. Thus, the water content should be quantified and/or adjusted to obtain reproducible NBT-based materials.

1.2.1.3 Layered Phases of Brownmillerite

When the perovskite layers are separated by some different motif, the perovskite derivatives will be formed as layered structures, also called as intergrowth. The Aurivillius phase consists of an intergrowth of perovskite layers with the $(Bi_2O_2)^{2+}$ layers, reported for the BIMEVOX compounds.

Layered perovskite derivative, $Ba_2In_2O_5$, possesses oxide-ion conduction. This material belongs to the brownmillerite type, which shows the general formula of $A_2B_2O_5$. It has the perovskite BO_6 layers alternated with BO_4 layers. The alternating tetrahedral–octahedral layers are formed using the ordering of intrinsic oxygen vacancy along [104] direction, and result in an orthorhombic cell.

It is reported that $Ba_2In_2O_5$ possesses oxide ion conductivity, with a transition from a low conductive phase to a high conductive one. The transition is related to conversion from orthorhombic structure to cubic phase above 925 °C, whereas the existence of the intermediate tetragonal phase is debated. The phase changes are concomitant by an anisotropic disordering of oxygen vacancies, which seems to involve only the equatorial oxygen atoms of the InO_4 tetrahedra. The oxygen vacancies here are also responsible for oxide-ion migration within the tetrahedral layers. Computational investigations also show the vacancies are not randomly distributed. The high-conductivities can keep stable at lower temperatures ranges using substituting or doping. The incorporation of PO_4^{3-} and SO_4^{2-} oxyanions make the subsequent enhancement of the low-temperature conductivity successful. Although some doped samples demonstrate good performances as SOFC electrolytes, these $Ba_2In_2O_5$-based ceramics are generally unstable under CO_2 and reducing atmospheres. Recently, $Ba_2In_2O_5$ has been studied for the possibility of as protonic conductors because of its good proton conductivity [48–50].

1.2.1.4 Ruddlesden–Popper Phases

Ruddlesden–Popper phases (structural archetype: K_2NiF_4) are considered particularly layered perovskites (with a chemical formula of $A_{n+1}B_nO_{3n+1}$). Such structures are composed of n perovskite layers alternated with rock-salt slabs, thus forming a framework in which n two-dimensional layers of BO_6 octahedra are joined along the stacking directions and separated by AO layers. Among the members of the Ruddlesde–Popper family, doped $Sr_3Ti_2O_7$ compounds were first reported exhibiting oxide ion conductivity. The doped compounds like Ti have been substituted by various elements that generally show p-type conduction at a high oxygen partial pressure and n-type conduction at lower oxygen partial pressure. The Ga- and the Co-doped samples exhibit the wider electrolytic window but both present very low conductivities.

Other ion conductors belonging to the Ruddlesden–Popper family are the oxides with the formula Ln_2NiO_δ (with Ln = La, Nd, Pr). These materials present one perovskite layer ($n = 1$) sandwiched between two rock-salt slabs. Depending on the oxygen nonstoichiometric, they generally crystallize either in the orthorhombic or in the tetragonal. The extra oxygen is accommodated at interstitial sites located on the rock-salt layers, where the interstitial sites are actively involved in the oxygen migration. Actually, experimental and molecular dynamics studies suggest that the oxide ion migration along with an interstitial mechanism. The vacant site is then filled by a second neighboring interstitial oxygen. The ionic diffusion is highly anisotropic in the *ab* plane. The Ln_2NiO_8-based compounds are mixed ionic and electronic conductors, which are mainly investigated for use as SOFC cathodes. The high oxide-ion conductivities reported among some doped compositions suggest that they can be used as electrolyte if the electronic contribution is suppressed [51–55].

Recently, oxide ion conductivity also has been reported for $NdBaInO_4$, which is a Ruddlesden–Popper derivative with a layered structure formed by perovskite units layers and NdO blocks layers (space group *P*21/*c*). This structure is peculiar because *A*-site cations (Nd and Ba) are ordered and the edge of the *B*-site octahedron (InO_6) faces the *A*O layer, in contrast with the common Ruddlesden–Popper structures in which the apical oxygen is the one facing the rock-salt unit [56–59].

Conductivity measurement under different oxygen partial pressures reveals that $NdBaInO_4$ exhibits oxide ion conductivity in the range 10^{-19} atm $\leq P(O_2) \leq 10^{-6}$ atm, respectively. The oxide ions diffusion is in the rock-salt layers in the *bc* plane. The oxide ion conductivity could be increased by substituting Nd with Sr. Oxygen vacancies generated due to substituting Sr, which are considered the charge carriers due to the induced increase of the ionic conductivity. It is partially replaced by various cations and the optimum is $NdBaIn_{0.8}Ti_{0.2}O_4$ in consideration of conductivity and electrolytic window. It suggests that the Ti-doped compounds with interstitial oxygen atoms have higher ionic conductivity compared with the parent compound [56, 58].

1.2.1.5 Hexagonal Derivatives

The principal structural motif of all the perovskite-based systems described before is the BO_6 octahedra connected by one or more of their vertices. However, it is possible to have perovskite structures in which the BO_6 octahedral are linked by their faces. In fact, when the cationic radius of *A* is particularly large (or the cationic radius of *B* is particularly small). The system possesses hexagonal symmetry when $t > 1$. These layers are the result of O_4 anionic layers in which a quarter of the anions have been replaced by *A* cation of similar sizes, such as alkaline and alkaline-earth metals. The stacking of these layers in the ideal perovskite is cubic type.

The hexagonal perovskite derivative of $Ba_3MoNbO_{8.5}$ has been reported by Fop in 2006. $Ba_3MoNbO_{8.5}$ consists of a hybrid structure within hexagonal perovskite blocks and palmierite blocks. $Ba_3MoNbO_{8.5}$ oxide has good stability and exhibits excellent oxide-ion conductivities in a wide $P(O_2)$ range. The high oxide-ion conductivities suggest that the hexagonal perovskites give new views on the design of novel oxide-ion conductors [60–62].

1.2.1.6 Aurivillius Phase of BIMEVOX Materials

BIMEVOX materials are famous oxide-ion conductors, based on the mother system of $Bi_4V_2O_{11}$. This oxide is an $n = 1$ member of the Aurivillius phases, where the general structure is composed of an intergrowth of n perovskite ($A_{n-1}B_nO_{3n+1}$) layers and bismuth oxide (Bi_2O_2) layers with Bi in square-based pyramidal coordination. The BIMEVOX compounds show high oxide-ion conductivities especially at moderate temperature ranges [63, 64]. Many metal cations could be introduced into the sites. However, because of their bad sensitivity to reduction atmosphere, the BIMEVOX compounds are difficult to use in SOFCs [63, 65, 66].

1.2.2 Properties of Dion–Jacobson-Type Layered Perovskites

The Dion–Jacobson-type layered perovskites form a homologous series with a general formula of A' [$A_{n-1}B_nO_{3n+1}$], where usually the $n > 2$. Similar to the layered perovskite of Aurivillius phases, the Dion–Jacobson phases do not link with a single perovskite phase because the perovskite-like layers such as $n \to \infty$ do not have independently. In an example like $CsLa_2Ti_2NbO_{10}$ in the Dion–Jacobson phase, the perovskite block layers are the La–O and Ti/Nb–O layers, but they do not consist of the $LaTiO_3$ perovskite or $LaNbO_3$ perovskite. $LaTiO_4$ and $LaTi_{0.5}Nb_{0.5}O_4$ compositions are allowed to show a different crystal structure. In fact, the Dion–Jacobson layered perovskites are similar to that of the A-site cations ordering double-layered perovskites and Ruddlesden–Popper layered perovskites.

$CsBiNb_2O_7$ and $RbBiNb_2O_7$ are well investigated as the Dion–Jacobson-type layered perovskites. They were reported by Subramanian et al. first, and characterized as orthorhombic. These compounds were identified with a larger unit cell and a lower symmetry structure. The neutron diffraction results show that $CsBiNb_2O_7$ is indeed polar. The first-principles calculations result also indicate the importance of octahedral rotations and cations displacements. Experiments also confirm that both of them belong to polar structures [67]. The polarity performances of $CsBiNb_2O_7$ and $RbBiNb_2O_7$ can be ascribed to the Bi^{3+} lone pair $6S^2$. But it cannot be used to explain why $CsNdNb_2O_7$ also shows a polar structure, where there are no lone pair cations. Similar to other layered oxides, the octahedral rotations have been optimized for the A-site coordination states and the octahedral rotations are also the reasons for the phase transition. On the other hand, same as the R–P phases, although the Dion–Jacobson-type layered perovskites are generally considered to have polar structure, the polar structure may be controlled in Dion–Jacobson-type layered perovskites through some other mechanism [68].

Polarization has been investigated in both $CsBiNb_2O_7$ and $RbBiNb_2O_7$ compounds and some experimental observation is generally in well agreement with the calculation prediction, and the number of polar Dion–Jacobson-type layered perovskites where the reported is growing. Furthermore, the alkali metal cations in the inter layers can make the materials susceptible to ion exchange like Li, H, Na, and water-uptake phenomena performance. Thus, further research should force on the compositions which have the Ba^{2+} in the A' sites of A' $[A_{n-1}B_nO_{3n+1}]$, such as $BaSrTa_2O_7$ and $BaTi_3O_{10,}$ and the previous studies indicate the nonpolar symmetry.

1.3 Crystal Structure Analysis

Understanding the structure of a certain system is important to truly understand and control the characteristics of a determined compound. Crystalline materials can be illustrated by an ordered arrangement of atoms in three dimensions. This arrangement constitutes a lattice, which can be reproduced by translation of an elementary parallelepiped containing the smallest group of atoms possessing the overall crystal symmetry. In general, the atomic distribution inside the unit cell (each of them individuated by three fractional coordinates x, y, z), and the symmetry (characterized by the space group) allows a better understanding of the crystal structure in the system [69–71].

Diffraction techniques offer a powerful method to characterize the crystal structure of many materials. The atomic species composing a certain structure can be imagined as arranged in periodic arrays defined by a set of planes called crystallographic planes. The sets of planes are described by the integer indices of h, k, and l. These (hkl) planes, and hence the atoms comprising them, form a diffracting grating able to scatter radiation of opportune wavelength, which is in the order of typical interatomic distances. A simple and efficient description of the diffraction from such planes is considered the diffraction as the consequence of simultaneous reflections of radiation by various lattice planes belonging to the same family. An incident radiation beam that makes an angle θ with the planes will produce a reflected wave with an equal angle θ. Since there are large numbers of parallel planes, reflections from successive planes will interfere with each other. In addition, there just be interference where the difference in path length between reflected waves of successive planes is a total number of wavelengths. This condition has been summarized using the Bragg's law

$$n\lambda = 2d_{hkl}\sin\theta \tag{1.3}$$

where n denotes an integer number known as an order of reflection (normally $n = 1$).

This relationship states that constructive interference between the scattered beams occurs only at certain values of θ, which are consequently associated with intense diffraction events generated by families of planes (hkl). If the angle of the incident radiation does not verify Bragg's law, diffraction will not have occurred due to the destructive interference.

The intensities of the diffracted signals are measured and used to obtain a diffraction pattern. In general, when the sample is considered a single crystal with a single continuous lattice, the crystal is rotated while irradiated by the incident radiation.

Rotation of the crystal allows each set of planes (*hkl*) to come into the correct orientation to satisfy Bragg's law and produce a diffraction signal. A single crystal diffraction pattern is in a way similar to a three-dimensional map composed of diffraction spots in the different directions, each of them carrying information about a specific set of planes. In the case of a powder composed of millions of crystallites randomly orientated, diffraction simultaneously occurs from every family of planes, which satisfies Bragg's law. All the planes (d_{hkl}-spacing) can be reflected at the same 2θ angle to the incident radiation and all these reflected beams will lie on a cone of semi-angle 2θ about the incident radiation. The plane gives rise to its own diffraction cone, of which base forms a ring (Debye–Scherrer ring) composed of a set of points, which corresponds to diffraction from a single crystallite. A detector can be rotated around a plane with an axis situated at the sample and intersect the Debye–Scherrer ring, thus producing a diffraction pattern in which the intensity is plotted with the Bragg angle. In this way, the three-dimensional structural information is spread on a one-dimensional pattern, increasing the complexity of the structural information extraction.

The intensity is related to a structure factor, $\boldsymbol{F}_{hkl}$, which describes the amplitude and phase of a wave diffracted from a certain family of planes (*hkl*). The structure factor is a quantity defined using the crystal structure, atoms states, the distribution in the lattice sites, occupancy of atoms and their thermal motion. $\boldsymbol{F}_{hkl}$ is mathematically expressed by

$$\mathrm{F}_{hkl} = \sum_{n}^{\mathrm{N}} g_n t_n f_n e^{2\pi i (hx_n + ky_n + lz_n)} \tag{1.4}$$

where n indicates the nth atoms in a unit cell and N denotes the total number of all the atoms. Occupation factor, g_n, is the fractional amount of the nth atom on a certain lattice site. t_n is the Debye–Weller factor, which considers the diminution in scattering caused by the thermal motion and displacement of the atoms, normally expressed by

$$t_n = e^{8\pi^2 U \frac{\sin^2\theta}{\lambda^2}} \tag{1.5}$$

where U is the atomic displacement factor, which in the isotropic approximation is related to the mean squared thermal displacement factor, u ($U = \langle u \rangle$).

A more accurate representation considers the anisotropy of the thermal displacement by describing atoms as ellipsoids defined by six different anisotropic displacement parameters, U_{ij}. Finally, the scattering factor, f_n, describes the interaction of the incident wave with a specific type of atom. The structure factors have a relation with the atomic distribution through a Fourier transform. Thus, the observed structure amplitudes, obtained from the diffracted signal intensities, are converted into a mathematical model representation of the atomic distribution.

As mentioned before, diffraction generates from a crystalline sample by illumination with a radiation of the opportune wavelength. X-rays have wavelengths comparable to the interatomic distances, which are suitable for the materials' structural study. Typical wavelengths used in X-ray diffraction (XRD) are in the range of ~0.5 to ~2.5 Å. Since X-rays are electromagnetic waves, they interact with the electronic cloud of the species composing the sample under investigation. In fact, the oscillating electric field of the incident X-ray radiation exerts a force on the electrons, forcing the electrons to oscillate with the same frequency as the electromagnetic wave. Electromagnetic radiation emits due to the acceleration and deceleration of the oscillating electron. Thus, it is clear how the ability of a certain atom to scatter X-rays is relative to the number of electrons, Z. This ability could be expressed using X-ray scattering factors, which depends on the electron distribution among the atomic nucleus and with the Bragg angle. Mathematically, the X-ray scattering factor (f_e) is expressed by

$$f_e(\mathbf{s}) = \int \rho_e(\mathbf{r}) e^{-2\pi i \mathbf{s}\mathbf{r}} d\mathbf{r} \tag{1.6}$$

where $\rho_e(\mathbf{r})$ is the electron density distribution of the atom measured at distance $\mathbf{r}$ apart from the center of atoms; $\mathbf{s}$ is the scattering vector, of which magnitude is equal to $2(\sin\theta)/\lambda$; and i represents the imaginary unit. With Z increasing, the electron density increases thus resulting in a greater scattering amplitude. It means that heavy atoms have better X-ray scatterers than light ones. Moreover, the scattering amplitude falls off with respect to $\sin\theta$, u to the interference of the radiation scattered from different electronic cloud parts.

Neutrons also are used to study crystalline materials by diffraction. The wavelengths of neutrons are determined by the De Broglie equation ($\lambda = h/(mv)$), which relates the mass (m) and velocity (v) of a free particle with the wavelength of the corresponding quantum–mechanical representation through the Planck constant, h. Neutrons with average thermal velocities ~2600 m/s are called thermal neutrons, which are suitable for diffraction studies with wavelengths ~1.5 Å. Different from X-rays diffraction, neutron diffraction is primarily diffracted using atomic nuclei. The efficacy of neutron scattering from a nucleus can be expressed by the neutron scattering length of the nucleus, b. The values of b do not vary systematically with Z, since the nucleus–neutron interaction is dependent on the details of the nuclear structure. b is simply expressed by

$$b = b_c + \frac{2b_i}{\sqrt{I(I+1)}} \mathrm{s} \cdot I \tag{1.7}$$

where coefficient b_c is the bound coherent, b_i is the incoherent scattering lengths, and s denotes the spin of the neutron and I is the modulus of angular momentum of the atomic nucleus.

Thus, the magnitude of the neutron scattering length varies greatly between neighboring elements and isotopes of the same element. The neutron scattering lengths are independent of the Bragg angle θ. As a result, the scattering of neutrons from an atomic nucleus is isotropic and the value of b does not change with the scattering angle. Because of the intrinsic natures in the X-ray and neutron scattering processes, these two diffraction techniques show evident differences. The interaction of neutrons is weaker than that of X-rays. Therefore, high neutron fluxes and a large amount of samples are necessary for typical neutron diffraction experiments. However, neutron diffraction has many important advantages. Since the values of b do not vary monotonically with Z, it is possible to distinguish atoms that have a similar number of electrons but different neutron scattering lengths. Light elements with sizeable neutron scattering lengths can be differentiated also in the heavy atoms. In fact, because of the large b of the oxygen atom, neutron diffraction allowed to get details of the oxygen even if alkaline-earth and transition metals were present.

1.3.1 Synchrotron X-ray Powder Diffraction

In the magnetic fields, the charged particles can receive a Lorentz force in a direction perpendicular to and direction of travel of these charged particles, thereby performing a circular motion. At this time, electromagnetic waves are generated in the tangential direction of the charged particles, where powerful continuous X-rays produced by electronic braking of the circular accelerator with this principle are called synchrotron radiation X-rays. Compared with the ordinary laboratory X-ray diffraction method, the biggest features of synchrotron X-ray diffraction are as shown in the following [72]. (1) High resolution and excellent in space-group determination and crystal system judgment. (2) High diffracted peak with high symmetry and strong directivity (such as Gaussian distribution). (3) High-order diffraction peak with high intensity due to the extremely high luminance. Measurement with a small amount of sample is available. (4) Wide energy range and arbitrary X-ray wavelength. (5) No peak splitting by $K_{\alpha 1}$ and $K_{\alpha 2}$. Thus, in comparison with the X-ray diffraction method, this method makes it possible to analyze the crystal structure with extremely high precision [72–74].

1.3.2 High-Temperature Synchrotron X-ray Powder Diffraction

The investigation of materials structures at high temperatures is very important. To study the structure at high temperature conveniently, the lattice constant could be investigated by in situ measurement with the X-ray diffraction method. More accurate temperature dependence of lattice constant and atomic variation parameter is able to obtain using the synchrotron X-ray powder diffraction method, which is an extremely powerful technique in accurately determining the crystal structure of the material at high temperatures [75, 76].

1.3.3 Neutron Diffraction

As described before, neutrons also can be used for diffraction measurements like X-rays. The advantages of using the neutron diffraction method for crystal structure analysis are as follows [74, 77]. (1) High scattering power from light elements. Thus, it is possible to specify the position of a light element, which is difficult by X-ray diffraction or electron diffraction. (2) Strong permeability and wide measurement possibility even at high-temperature ranges and/or high pressures. (3) High intensity at a low lattice spacing region and accurate atomic displacement parameters. (4) Possibility to distinguish adjacent or nearby elements of atomic number [78–80].

Neutron diffraction is used for crystal structure studies, where only the X-ray diffraction method is not enough, especially in the use to locate light atoms, such as hydrogen and oxygen. Usually, the X-ray method solves the main part of the structure, whereas neutron diffraction locates the light atoms. Neuron diffraction is also used to distinguish atoms that have similar X-ray scattering factors, such as Mn, Fe, Co, and Ni. There are several characteristic differences between X-ray diffraction and neutron diffraction methods. First, neutron-scattering lengths of any atoms are quite different from those of the X-ray atomic scattering factors. These factors are simple functions of atomic numbers. On the other hand, there is no simple dependence of neutron scattering length on atomic number. Therefore, light atoms like hydrogen and oxygen give very weak X-ray diffractions, but strong neutron diffractions. Thus, the precision of atomic coordinates for light atoms such as O and H obtained from neutron diffraction experiments is higher than those from X-ray diffraction. Another difference is that neutrons give a continuous spectrum of radiation instead of intense characteristic peaks by X-ray diffraction.

Furthermore, there are some technical advantages in measuring neutron diffraction compared with X-ray measurement:

1. Absorption coefficients for neutron are generally lower, thus larger samples can be used that give a better powder average during data collection and the preferred orientation is less. Samples can be measured under various environments such as heating, cooling, high-pressure, magnetic field, etc.
2. Neutron diffraction gives the information of bulk, while X-ray diffraction gives the information of the surface. However, the surface is easily reconstructed (sintering, preferred orientation changes, etc.) at high temperatures, thus it is often difficult to get satisfactory data for Rietveld analysis from X-ray diffraction.
3. The peak shape functions are generally simpler in neutron diffraction patterns compared with those of conventional X-ray diffraction patterns.

4. Neutron scattering lengths don't have essentially angular dependence in the range of interest, so meaningful data can be collected at much higher angles and the thermal parameters can be estimated much more precisely.

Neutron diffraction technique gives important information that cannot be obtained from other diffraction techniques as mentioned above. Ceramics as perovskite-related compounds are key materials for the current science and technology. The property of ceramic materials has a strong correlation with existing phases and crystal structure, thus many researchers study the crystal structure by conventional X-ray diffraction. Usually, ceramic materials are fabricated at high temperatures above 800 °C, and many ceramic products are utilized at high temperatures. Thus, in situ neutron diffraction techniques at high temperatures are really necessary to establish the structure–property correlation [81].

1.3.4 High-Temperature Neutron Diffraction

The advantages of using the neutron powder diffraction method for in situ measurement at high temperature are mainly as described below [79, 82, 83]. (1) Bragg Brentano optics are used in high-temperature X-ray powder diffraction measurements. Thus, to obtain good quality data, it is important to make a tabular sample surface smooth and the crystal orientation on the surface averaged and disordered. However, at high temperatures, the surface of the sample is not smooth due to grain growth, sintering, evaporation, chemical reaction, difference in thermal expansion from the sample holder, etc., and orientation problems often occur. Whereas, neutrons have higher permeability than X-rays. Therefore, in neutron diffraction, it is possible to measure powder samples in the container about 1 cm in diameter or the rod sample with a diameter of about 1 cm by the transmission method. In X-ray diffraction, a signal is obtained from the vicinity of the surface, whereas neutron diffraction obtains bulk information inside the sample. Accordingly, it is possible to greatly reduce the problems of orientation and coarse particles, which frequently occur by X-ray powder diffraction. This is a great advantage in the refinement of the crystal structure by Rietveld analyses. In addition, changes in profile due to contamination and alteration of the surface can be avoided. It is less susceptible to the influence of oxides on the surface generated when measuring non-oxide at high temperatures. (2) In an X-ray diffraction measurement, the details about the sample surface and the position of the diffraction peak change when the height of the surface changes due to thermal expansion. Whereas, because neutron diffraction uses a large sample, the position of the diffraction peak is less susceptible to thermal expansion. Thus, lattice constant at different temperatures can be obtained accurately. (3) When the light element is deeply involved in the structural change and phase transition of the material occurring at high temperatures, the behavior cannot be accurately grasped by X-ray measurement. However, by using neutron diffraction with high scattering power from a light element, it is possible to accurately grasp its behavior, investigate the details of structural change, and clarify the phase transition mechanism [78, 82, 84–86].

1.3.5 Rietveld Refinement

In a powder diffraction pattern, the three-dimensional (3D) structure of a determined system is projected into a two-dimensional (2D) space. Therefore, it inevitably leads to some loss of information about the crystal structure, principally caused by the overlap of the various reflections. The Rietveld method can be used to extract detailed structural information from a powder diffraction pattern. This method is a least-squares refinement process, which involves the refinement of various parameters of an initial structural model to minimize the different points between the experimental diffraction pattern and the calculated diffraction pattern obtained from the refined model [73, 81, 83, 87–89].

The general Rietveld procedure uses the observed profile intensities of the diffraction pattern to extract the maximum amount of information. Due to the overlap of reflections, the measured intensity, y_i^{obs}, at a certain point i is defined as

$$y_i^{obs} = y_b^{obs} + \sum_i y_i^{obs}(\text{Bragg}) \quad (1.8)$$

where ${y_i}^{obs}$ is the background intensity and the summation of y_i^{obs} (Bragg) is the contribution of all the Bragg reflections close to step i. The fitting between the observed experimental pattern and the refined calculated pattern is got by minimizing the residual S_y, which is the squared difference in the observed data and the calculated data at each data point

$$S_y = \sum_i w_i \left[y_i^{obs} - y_i^{cal}\right]^2 \quad (1.9)$$

where $w_i = 1/y_i^{obs}$ is an opportune weight factor and y_i^{cal} is the intensities of calculated diffraction pattern at ith point. And the calculated intensity includes several factors and is defined by the summation performed on each hkl reflection:

$$y_i^{cal} = s\sum_{hkl} |F_{hkl}|^2 \phi_{hkl} L_{hkl} P_{hkl} A + y_b^{cal} \quad (1.10)$$

where s is a scale-factor; $\boldsymbol{\phi}_{\text{hkl}}$ is the reflection profile function; the term $\boldsymbol{L}_{hkl}$ contains the Lorentz factor (correction for the θ dependence of integrated intensities), the polarization factor (arising from partial polarization of the scattered waves), and the multiplicity factor (accounting for the multiplicity of a certain reflection); $\boldsymbol{P}_{hkl}$ denotes the orientation function, and A is an absorption factor; y_b^{cal} is an intensity of the calculated background.

The structure factor contains all the information about the atomic positions and the thermal and occupancy parameters, and is correlated to the scattering density via a Fourier transform.

The profile function is used to model the shape of a particular Bragg reflection. Generally, a pseudo-Voigt function is employed, that is a function using Gaussian and Lorentzian:

$$\phi_{hkl} = \eta L + (1 - \eta)G \tag{1.11}$$

where η is the pseudo-Voigt mixing parameter, indicating the fraction of Lorentzian (L) and Gaussian (G) components. It is influenced by instrument and specimen factors and depends on the displacement of the reflections from the profile positions and width coefficients. The match between the observed pattern and the calculated one is dependent on the unit cell parameters and the 2θ correction. They are often highly correlated but should be determined independently. For angle dispersive data, the variation of full width at half maximum (FWHM) as function θ of the Gaussian (G_{FWHM}) and Lorentzian (L_{FWHM}) parts is usually modeled by

$$G_{\text{FWHM}} = \left[U \tan^2\theta + V \tan\theta| + W + \frac{Z}{\cos^2\theta} \right]^{1/2} \tag{1.12}$$

$$L_{\text{FWHM}} = X \tan\theta + \frac{Y}{\cos\theta} \tag{1.13}$$

where U, V, and W are parameters, Z is the Scherrer coefficient for Gaussian particle size broading, and X and Y model the Lorentzian strain and broadening, respectively. For TOF neutron diffraction data, the general pseudo-Voight function has to be modified to consider the profile asymmetry induced by the neutron pulse. Actually, the neutron pulse is asymmetric in time: the leading edge is sharp, while the trailing part decays according to the temperature and moderator size. This asymmetry is generated by convolution of the pseudo-Voigt function with separate raising and falling exponentials that describe the time dependency of the pulse:

$$\phi_{hkl}^{\text{TOF}}(t) = \int \phi_{hkl}(\Delta\text{T} - t)\text{I}(t)\text{d}t \tag{1.14}$$

where $\phi_{hkl}^{\text{TOF}}(t)$ is the TOF profile function, $\phi_{hkl}(\Delta T - t)$ is the time-dependent pseudo-Voigt function, ΔT is the difference in the time-of-flight between reflection positions and that of profile points, and $I(t)$ is the function describing the time dependency of the pulse. Depending on the value of t, $I(t)$ rise or decay exponentially:

$$I(t) = Ne^{at^*} \quad t^* < 0; \quad I(t)'' = Ne^{-\beta t^*} \quad t^* > 0 \tag{1.15}$$

where N denotes the normalization factor, t^* is a reference time corresponding to the maximum pulse, and α and β are the rise and decay constants, respectively. The variation of the FWHM is given by complex polynomial equations of the Gaussian and Lorentzian widths and the d_{hkl}-spacing value of the determined reflection.

The background signal in a powder diffraction experiment occurs due to several factors such as thermal diffuse scattering, incoherent scattering, Compton scattering, parasitic scattering from the sample container, sample environment, and amorphous content in the sample. The background can be either linearly interpolated and subtracted or fitted using a polynomial or semi-empirical function containing refinable parameters.

Generally, the Rietveld algorithm involves the solution of equations given by the derivatives using the calculated results with respect to those of various parameters, without going into the mathematical details. These refinable parameters define the structural model (cell dimensions, atomic positions, atomic displacement parameters, fractional occupancies) and the fitting profile (background and profile parameters, instrumental features, scale factors). The solutions to the set of normal equations are found with a procedure, in which the calculated shifts of the refined parameters are used in the initial parameters to crease better models. The procedures are reiterated until the residual equation is minimized. Various statistical parameters can be used to assess the quality of the employed model and monitor the refinement. These reliability factors (R factors) are defined as

$$R_p = \frac{\sum_i y_i^{obs} - y_i^{cal}}{\sum_i y_i^{obs}} \tag{1.16}$$

$$wR_p = \left[\frac{S_y}{\sum_i w_i \left(y_i^{obs}\right)^2}\right]^{1/2} \tag{1.17}$$

Particularly, wR_p is the weighted profile R-value, representing the weighted sum of the difference between observed results and estimated data; it is the most significant indicator of the quality of fit as directly contains the quantity being minimized during the refinement, S_y. The statistically expected R-value (R_{exp}) can be calculated as

$$R_{exp} = \left[\frac{N - P}{\sum_i w_i \left(y_i^{obs}\right)^2}\right]^{1/2} \tag{1.18}$$

where N denotes the number of observations and P denotes the refined parameters. R_{exp} is used to define goodness-of-fit (χ^2), which should be as close as possible to the unity.

$$\chi^2 = \frac{wR_p}{R_{\text{exp}}} \tag{1.19}$$

Although the above statistical parameters give a good quality indication of a certain Rietveld fit, their values are sometimes misleading as they are affected by various factors such as counting time, statistics, and the background intensity. Thus, the quality of the refinement also should be monitored by checking the profile fit, the nature of the shifts/estimated standard deviations (esd) of the different refined parameters, and the plausibility of the structural model. Rietveld refinements using the neutron diffraction method for various samples were performed using the Z-code software [83].

1.4 Electrical Properties Measurement

The total electrically charged particles in oxides are generally electrons and or ions. The ion carriers usually consist of cations like Li^+, Na^+, and K^+, anions like O^{2-}, F^-, and foreign ions like impurity ions, dopant ions, and protons. And the electron carriers are generally electrons and electron holes.

The electrons' and electron holes' mobility is much higher than that of ions, most of the oxides are electron conductors. Normally, only one kind of charge carrier predominates under different conditions (temperature and PO_2). In electron-conducting oxide, it will be an n-type conductor when electrons transport predominates and will be a p-type conductor when electron holes predominate. On the other hand, parts of oxides may be ionic conductors or mixed ionic electronic conductors at different atmospheres. Some oxides exhibit proton conduction under hydrogen and or water atmospheres. The SI unit used in the conduction is Siemens per cm (S cm^{-1}). The charge carriers in binary oxide materials are cations, anions, and electrons, where both of them have the ability to show different transport carries. Usually, the most important contributions to conductivity are defect electrons (n-type conduction) and electron holes (p-type conduction). Therefore, the total electrical conductivities can be given as the following:

$$\sigma = \sigma_{\text{ion}} + \sigma_{\text{el}} = \sigma_c + \sigma_a + \sigma_n + \sigma_p \tag{1.20}$$

whereσ_c, σ_a, σ_n, and σ_p denote the cationic, anionic, defect electron, and electron holes conductivities, respectively.

The ionic conduction (cations and anions) may contribute to vacancy and or interstitial cations. Generally, just one kind of carrier dominates the charge transport in an oxide. Therefore, contributions from other minority carriers can be neglected as an approximation in many cases. For oxides, because the electrons and electron holes are usually several orders of magnitude higher than those of the diffusion of the ions. Thus even if the electron concentrations are lower than that of the ion charge carriers, the oxide will still be an electron conductor. The relative concentrations of ion and electron often vary with temperatures and or oxygen pressures. The particles with charge $Z_i e$ can be shown in terms of charge carriers' mobilities or conductivities as follows.

$$j_i = z_1 e c_i B_i E = c_i u_i E = \frac{\sigma_i E}{z_i e} \tag{1.21}$$

$$j_i = D_i \frac{c_i Z_i e E}{kT} \tag{1.22}$$

$$D_i = B_i kT = u_i \frac{kT}{z_i e} = \sigma_i \frac{kT}{c_i z_i^2 e^2} \tag{1.23}$$

$$\sigma_i = (Z_i e)^2 c_i D_i / kT. \tag{1.24}$$

1.4.1 Temperature Dependence of Total Electrical Conductivities

Generally, the two-terminal method and four-terminal method are used for measurements related to electrical conduction. The two-terminal method is the simplest method for conductivity measurement, however, the effect such as the interface between the electrode and the pellets leads to the resistance value obtained from the two-terminal method. Thus, the two-terminal method is used for high resistivity materials such as organic crystals and macromolecules, due to the little influence of contact resistance. Whereas the four-terminal method is used as a standard measurement method of resistivity measurement, where measurement performs without the influence of contact resistance which is different from the two-terminal method. When a constant current I flows through the measurement cell, voltage V is applied in a certain length d as follows:

$$\sigma = \frac{I \times d}{V \times S} \tag{1.25}$$

where S is cross-section area. The segments are uniform; the strength of the electric field is not so large. σ is electrical conductivity, which is determined by the kind and state (temperature, pressure, etc.) of the material.

In the mixed conductors, high-temperature conductivities are generally attributed to electrons, electron holes, and oxide ions. By examining the oxygen partial pressure (PO_2) dependence of total electrical conductivities, we can evaluate ion, hole, and electron conduction.

1.4.2 The Oxygen Partial Pressure Dependence of Total Electrical Conductivities

In general, oxygen in the gas phase dissolves into oxygen vacancy at a higher oxygen partial pressure range. The equation is given as follows with Kröger–Vink notation.

$$\frac{1}{2}\mathrm{O}_2(g) + \mathrm{O}_\mathrm{O}^\times V_O^{\cdot\cdot} \Leftrightarrow O_O^\times + 2h^\cdot \tag{1.26}$$

where $V_O^{\cdot\cdot}$, $O_O^\times$ and $h^\cdot$ are lattice vacancy of oxygen with positive divalent charge, oxygen in lattice, electronic hole, respectively. If oxygen behaves as an ideal gas, the reaction rate constant K is given as below.

$$K = \frac{a_{\underset{O}{O\times}} a_{h^{\cdot\cdot}}^2}{a_{\mathrm{O}_2}^{1/2}(g) a_{V_O^{\cdot\cdot}}} = \frac{\frac{[\underset{O}{O\times}]}{O}\left(\frac{n_p}{N_V}\right)^2}{\left(\frac{P(\mathrm{O}_2^0)}{P(\mathrm{O}_2)}\right)^{1/2}\frac{[V_O^{\cdot\cdot}]}{O}} \frac{[\underset{O}{O\times}]}{[V_O^{\cdot\cdot}]}\left(\frac{n_p}{N_V}\right)^2\left(\frac{P(\mathrm{O}_2^0)}{P(\mathrm{O}_2)}\right)^{1/2} \tag{1.27}$$

where n_p is the hole concentration, and N_v is the density of state of the valence band. $p_{\mathrm{O}_2^0} = 1\,\mathrm{atm}$. Then K' is expressed as follows when $[O_O^\times] = 1$

$$K' = \frac{n_p^2}{[V_O^{\cdot\cdot}]P(\mathrm{O}_2)^{\frac{1}{2}}} \tag{1.28}$$

$$n_p = \left(K'[V_O^{\cdot\cdot}]\right)^{\frac{1}{2}} P(\mathrm{O}_2)^{\frac{1}{4}} \tag{1.29}$$

When the change in the oxygen vacancy concentration is negligibly small at constant temperature, the relationship between the concentration of holes and $P(\mathrm{O}_2)$ can be explained with the following relation:

$$[h^\cdot] \propto P(\mathrm{O}_2)^{\frac{1}{4}} \tag{1.30}$$

Since the structure does not change significantly, it is considered that the carrier mobility does not change. When the dominant conduction carrier is a hole, the conductivity is denoted as the following equation:

$$\sigma_{\text{total}} \approx [h^{\cdot}] e\mu_h \tag{1.31}$$

Therefore,

$$\sigma_{\text{total}} \propto P(O_2)^{\frac{1}{6}} \tag{1.32}$$

Under low oxygen partial pressure, oxygen atoms in the lattice form the oxygen gas and the oxygen vacancy,

$$O_O^{\times} \Longleftrightarrow \frac{1}{2}O_2(g) + V_O^{\times} \tag{1.33}$$

$$K' = \frac{[V_O^{\times}]}{[O_O^{\times}]} P(O_2)^{\frac{1}{2}} = [V_O^{\times}] P(O_2)^{\frac{1}{2}} \tag{1.34}$$

Then oxygen vacancy provides electrons as follows:

$$V_O^{\times} \Leftrightarrow V_O^{\cdot} + e' \tag{1.35}$$

$$K_{V1} = \frac{[V_O^{\cdot}]}{V_O^{\times}} n_e \tag{1.36}$$

Second, if oxygen vacancy provides more electrons

$$V_O^{\cdot} \Leftrightarrow V_O^{\cdot\cdot} + e' \tag{1.37}$$

$$K_{V2} = \frac{[V_O^{\cdot\cdot}]}{[V_O^{\cdot}]} n_e \tag{1.38}$$

From Eqs. (1.34), (1.36), and (1.38), the following equations are obtained:

$$[V_O^{\times}] = \frac{[V_O^{\cdot\cdot}]}{K_{V1}K_{V2}} n_e^2 \tag{1.39}$$

$$[V_O^{\cdot\cdot}] = \frac{K' K_{V1} K_{V2}}{n_e^2 P(O_2)^{\frac{1}{2}}} \tag{1.40}$$

$$[V_O^{\cdot}] = \frac{K' K_{V1}}{n_e^2 P(O_2)^{\frac{1}{2}}} \tag{1.41}$$

To satisfy the charge valance condition, the following equation is obtained:

$$n_e = [V_O^{\cdot}] + 2[V_O^{\cdot\cdot}] \tag{1.42}$$

From Eqs. (1.40) to (1.42), the following equations are obtained:

$$n_e^3 = K' K_{V1}(2K_{V2} + n_e)P(O_2)^{-\frac{1}{2}} \quad (1.43)$$

When $n_e << 2K_{V2}$,

$$n_e - (K' K_{V1} 2K_{V2})^{1/3} P(O_2)^{-\frac{1}{6}} \quad (1.44)$$

$$\sigma_{total} \propto P(O_2)^{-\frac{1}{6}} \quad (1.45)$$

When $n_e >> 2K_{V2}$ (dominant carrier in conduction is electron) [90],

$$n_e = (K' K_{V1})^{1/2} P(O_2)^{-\frac{1}{4}} \quad (1.46)$$

$$\sigma_{total} \propto P(O_2)^{-\frac{1}{4}} \quad (1.47)$$

Generally, the conductivity is not depending on $P(O_2)^{-1/4}$ when the dominant carrier is an oxide ion. The diffusion coefficients for oxygen ions can be predicted from the conductivity measurement using the Nernst–Einstein relation. The Nernst–Einstein relationship is used to obtain a qualitative measure of the transfer number for the material. The transport number is obtained using comparing the ion conductivities and the calculated conductivities derived from the ionic diffusion coefficient, as follows:

$$\sigma_{ion} = \frac{(Z_i e)^2 C_{ion} D_{ion}}{kT} \quad (1.48)$$

where Z_i is the effective charge of each ion, e is the charge of the electron, and C_{ion} is the total concentration of all the ions.

1.4.3 Oxygen Concentration Cell Measurements

The total electrical conductivities can be regarded as the sum of the partial conductivity under the charge carriers:

$$\sigma = \sum_i \sigma_i \quad (1.49)$$

Ionic transport number (or transference number) measurements can be used to assess the ability of a certain system to support ionic conduction. The ionic transport number (t_{ionic}) corresponds to the fraction of the total current carried in a solid electrolyte by a given ionic species. The transport number t_i of species i denotes the ratio of the partial conductivity to the total conductivities

$$t_i = \frac{\sigma_i}{\sigma} \tag{1.50}$$

$$\sigma = \sigma_{\text{ion}} + \sigma_{el} = \sigma_c + \sigma_a + \sigma_n + \sigma_p \tag{1.51}$$

$$\sigma = \sigma(t_{\text{ion}} + t_{el}) = \sigma(t_c + t_a + t_n + t_p) \tag{1.52}$$

A good ionic conductor should ideally have a t_{ionic} value close to unity. For a solid oxide electrolyte, the t_{ionic} value is used to reveal the nature of the observed conductivity. The electromotive force measurements using a concentration cell can determine the ion transport number of a certain material. When an oxide is subjected to an oxygen gradient, an emf voltage is produced over the sample. This voltage is calculated with the following equation, assuming that the ionic transport number associated with the oxygen flow is constant within the gradient

$$\text{E}_{\text{obs}} = \text{t}_{\text{ionic}}\left[\frac{-\text{RT}}{nF}\ln\left(\frac{a^{\text{I}}}{a^{\text{II}}}\right)\right] \tag{1.53}$$

where E_{obs} is the observed voltage, T is the temperature, R is the ideal gas constant, n is the number of electrons, F is the Faraday's constant, and a^1 and a^2 are the oxygen activities at the two sides of the oxygen gradient.

Experimentally, the oxygen activity gradient is controlled by using a gradient of oxygen partial pressure $P(O_2)$. When an electrolytic material sample, with applied reversible electrodes, is held between two different oxygen partial pressures ($p_{O_2}{}^{\text{I}}$ and $p_{O_2}{}^{\text{II}}$), the resulting Nernst equation is

$$\text{E}_{\text{Nernst}} = \frac{-\text{RT}}{4F}\ln\left(\frac{p\text{O}_2^{\text{I}}}{p\text{O}_2^{\text{II}}}\right) \tag{1.54}$$

Therefore, it can be derived that

$$\text{t}_{\text{ionic}} = \frac{E_{\text{obs}}}{E_{\text{Nernst}}} \tag{1.55}$$

The ionic transport number is determined by simply measuring the open-circuit voltage (E_{obs}) generated across the electrolytic sample under the $P(O_2)$ gradient. The $P(O_2)$ pressure gradient is generated by mounting the specimen between two separated chambers, in each of which regulation of the type and the gas flow is possible. The use of different couples of gases for control of the temperature allows the determination of t_{ionic} under various conditions.

The ionic transport number allows to discriminate whether a material is an ionic conductor or not. However, especially when the material is used, it is also important to individuate the electrolytic domain of the solid conductor. The electrolytic domain is generally defined as the $P(O_2)$ range, in which the conductivity of the system is purely ionic. ##Since ion conductivity in oxides generated are caused by the oxygen defects such as oxygen vacancies, the $P(O_2)$ dependence of the conductivities can be used to explain employing some defect equilibrium reactions. It is possible to assume that in an undoped sample, the introduction of oxygen defects is the normal result of the interaction between the oxide and the oxygen in the surrounding atmosphere. Thus, only the oxygen addition and removal reactions are considered.

1.4.4 Impedance Spectroscopy Measurements

Impedance spectroscopy measurement is a method to determine the electrical properties of a ceramic material. This method is particularly useful as it can be used for the separation of the electrical responses of the different regions of the specimen (such as bulk, grain boundaries, and electrode interfaces). Impedance measurements are performed over a frequency range and the different electrical responses are differentiated according to the relaxation times of the various material regions. Different from the direct current (D.C.) techniques, impedance spectroscopy uses an alternating current/voltage source that induces a sinusoidal variation of the voltage (V) and current (I) with time; impedance spectroscopy is, therefore, an alternating current (A.C.) technique [21, 26, 91, 92].

When a system is under a DC voltage, the resistance is defined by the well-known Ohm's law, $R = V/I$. However, in the case of an ACvoltage, the electric charge flow alternatively changes direction with time, introducing a similar temporal variation in the induced current. The sinusoidal AC voltage and the generated current can be described by

$$V(\omega) = V_0 \sin(\omega t) \tag{1.56}$$

$$I(\omega) = I_0 \sin(\omega t + \phi) \tag{1.57}$$

where V_0 and I_0 are the potential and current value at $t = 0$; ω is the radial frequency given by $\omega = 2\pi f$ (with f indicating the frequency); ϕ is phase shift (current with respect to the applied voltage). In such a system, the opposition to the current flow is defined by the complex impedance Z^*:

$$Z^* = \frac{V(\omega)}{I(\omega)} = Z_0 e^{j\phi} = Z' + JZ'' \tag{1.58}$$

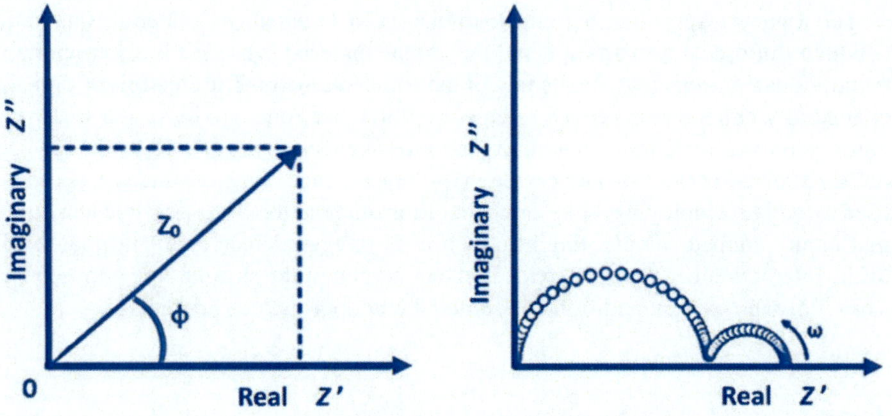

Fig. 1.8 Vector representation of complex impedance and typical complex impedance plot diagrams. Z^* is a vector quantity, possessing both magnitude and phase, and can be expressed by complex numbers through the sum of a real part, Z' and an imaginary part Z''. Ideal complex impedance representation of two RC elements in series

where j denotes the imaginary unit. $\mathbf{Z}^*$ is a vector quantity, possessing both magnitude and phase, which is expressed by complex numbers through a real part (Z') and an imaginary part (Z'') (Fig. 1.8). The real component is associated with the resistance of the system while the imaginary component is related to the reactance of the system. The reactance component contains capacitive (charge storage induced by the alternating voltage) and inductive (electromotive force generated by the change of the current flow) effects produced by the system's response to the sinusoidal voltage.

$$Z'(\omega) = Z_\infty + \frac{(Z_0 - Z_\infty)\left[1 + (\omega\tau)^\beta \cos\frac{\pi}{2}\beta\right]}{\left[1 + (\omega\tau)^\beta \cos\frac{\pi}{2}\beta\right]^2 + \left[(\omega\tau)^\beta \sin\frac{\pi}{2}\beta\right]^2} \tag{1.59}$$

$$Z''(\omega) = -\frac{(Z_0 - Z_\infty)(\omega\tau)^\beta \sin\frac{\pi}{2}\beta}{\left[1 + (\omega\tau)^\beta \cos\frac{\pi}{2}\beta\right]^2 + \left[(\omega\tau)^\beta \sin\frac{\pi}{2}\beta\right]^2} \tag{1.60}$$

Impedance data is represented in a complex plane diagram, which plots the imaginary part Z'' versus the real part Z'. Generally, data has the shape of a semicircle, of which each point corresponds to a different measured frequency in a typical complex plane diagram. The semicircle shape gives information about the electrical relaxation type. If the semicircle is a perfect arc, for example, the impedance response corresponds to a single and well-defined relaxation process (normally ascribed to an ideal behavior). A depressed semicircle is an indication of a distribution of relaxation times, while multiple semicircles (which can be more or less resolved) are proof of multiple electrical responses. The complex impedance plots are helpful in highlighting the most resistive components of a material and are used to extract information about the resistance (R) and capacitance (C) values associated with each electrical response. The intercept arc in the real axis denotes the resistance of the electrical response associated with the semicircular signal, whereas the capacitance value can be estimated from the frequency at the semicircle's maximum ($C = 1/(\omega_{max}R)$). The capacitance values are particularly useful in identifying which area of the sample is responsible for a certain electrical response. In fact, the capacitance value of a microstructural region depends on the thickness of the region itself, so different regions have different capacitance values. The crystal grains (bulk) of the specimen present a smaller capacitance (about 10^{-12} F) than the grain boundary (from 10^{-11} to 10^{-8} F), while for the electrode interface, large capacitance values about from 10^{-7} to 10^{-5} F are usually associated.

As explained before, the analysis of complex impedance plots allows the extraction of resistance and capacitance values for the various electrical responses. However, a more accurate interpretation would make use of an equivalent circuit fitting procedure. In such a procedure, an equivalent circuit is built using various electrical elements in series and/or parallel combinations. The equivalent circuits are utilized to refine the impedance data and extract more accurate values for the different quantities. The chosen model must have a physical basis on the electrochemistry of the system and should be formed by the smallest possible number of elements. Without considering the detail, the fitting method is a complex nonlinear least-square procedure, in which the parameters defining the equivalent circuit are reiteratively refined until the best fit between the calculated and the observed spectra is obtained. The initial parameters (such as R and C values) are manually extracted by examination of the complex impedance plot.

Typical impedance data obtained from conducting electro ceramics is modeled by series and parallel combinations of R and C elements. Usually, polycrystalline solid electrolytes are represented using a series array of R–C elements, where R–C denotes a resistor in parallel with a capacitor. The impedance values for a resistor (Z_R) and a capacitor (Z_C) are given by

$$z_R = R \tag{1.61}$$

$$z_c = \frac{1}{j\omega C} \tag{1.62}$$

Normally, C is used only when a certain semicircle shows an ideal shape, i.e. when the electrical relaxation is well defined and related to only one relaxation time. However, the electrical response of a real material generally presents a distribution of relaxation times. In fact, in polycrystalline materials, the presence of grains of various sizes and shapes, grain boundaries, porosity, inhomogeneities, surface roughness at the sample–electrode interface results in distributed electrical responses. As mentioned before, when the semicircle is depressed, the relaxation time associated with the electrical response is not a single value but is distributed around a mean one. The ideal capacitor (C) is replaced by a constant phase element (CPE), which works particularly well in modeling distributed responses. The impedance of a CPE is given by

$$Z_{CPE} = \frac{1}{(j\omega)^n Q} \tag{1.63}$$

where n is an exponential factor that quantifies the departure from ideality ($n = 1$ for the ideal capacitor) and Q is a factor expressed in S rad-n. The capacitance value of a resistor in parallel to a CPE (C_{CPE}) can be calculated by

$$C_{CPE} = \left[QR^{(1-n)}\right]^{1/n} \tag{1.64}$$

where C_{CPE} is used to assess the region of the material generating that particular electrical response in the same way as described for the value of the ideal capacitance.

The total impedance of a certain equivalent circuit configuration is calculated from the impedance values of the various series and/or parallel elements. In fact, the impedance values for combinations of elements in series (Z_{series}) and in parallel ($Z_{parallel}$) are given by the simple relationships:

$$Z_{series} = \sum_i z_i \tag{1.65}$$

$$Z_{parallel} = \left[\sum_i \frac{1}{Z_i}\right]^{-1} \tag{1.66}$$

where Z_i denotes the impedance of the ith element.

Impedance spectroscopy allows extraction of the resistance values for various electrical relaxation phenomena and, therefore, is used to characterize the conductivity (σ) of various parts composing the microstructure of the system. The conductance values (G) are easily obtained from the various resistances ($G = 1/R$). The conductance is related to the conductivity value by the geometric factor of the material, $\sigma = (l/A)G = 1/\rho$, where l is the length of the measured sample, A is the cross-sectional area of the sample, and ρ is the resistivity value. For ionic conductive materials, variable temperature (T) impedance spectroscopy measurements are particularly useful, since the conductivity of these systems is thermally activated. The following equation shows the Arrhenius relationship:

$$\sigma = Ae^{-E_a/k_B T} \tag{1.67}$$

where A is a pre-exponential factor, E_a is the activation energy, and k_B is the Boltzmann's constant. It allows to produce plots of ln(σ) versus $1/T$, which highlights the thermal dependence of the conductivity directly. In this representation, the conductivity data are presented as lines of which slope corresponds to the activation energy of the particular conduction process. Changes in the slope of a certain line indicate the change in the conduction process associated with possible phase transitions or structural modifications.

In typical experiments, the impedance spectroscopy is measured with a single-frequency AC voltage under investigation and to measure the phase shift and amplitude of the resulting current at the given frequency. The frequency is swept in a certain range and all the data points are collected at different frequencies. Various measuring instruments are employed, where the most common one being the frequency–response analyzer. A typical frequency–response analyzer determines the impedance by correlating the sample response with two synchronous signals, one in phase with the applied voltage and the other phase-shifted of 90°. It allows obtaining the in-phase and out-of-phase components of the impedance signal at each frequency. Frequency–response analyzer normally possesses an auto integration feature that averages the result over a predetermined number of cycles to produce statistically consistent results.

In the general procedure employed for the impedance spectroscopy experiments, a synthesized pellet with about 20 mm in diameter and 2 mm in thickness is used, which is sintered at a temperature, which was determined by the different samples. Pt electrodes or Au electrodes are painted on both sides of the pellet using a Pt-paste or Au-paste and successively fired to remove the organic solvent at 1000 °C and harden the metal electrodes. The sample with applied Pt electrodes is then loaded in a homemade impedance jig. The impedance jig is constituted by a mullite tube with a sample chamber at one of its ends; the test sample is inserted in the chamber, sandwiched between two Pt wires, and held in position by some ceramic discs acting as a spring. A thermocouple is set nearby the pellet to monitor the accurate temperature inside the chamber. The impedance jig is fitted inside a sealed tube furnace, in which the temperature and the gaseous atmosphere are controlled. Impedance spectra for all the samples were recorded with a Solartron 1260 impedance analyzer in the frequency range 0.1–15 MHz with an applied alternating voltage of 0.1 or 0.4 V. Impedance spectroscopy was recorded under various gases flow (regulated with a flow meter). Dry and moist gas flows were produced by, respectively, flowing the gas in a Dreschel tube filled with some desiccant or water. Data were collected at various temperatures, allowing 1 h for the sample to equilibrate under the atmosphere. The obtained impedance spectroscopies were corrected by an analyzer and analyzed using the ZView software.

1.5 Bond-Valence-Based Energy Calculation

Bond-valence (BV) method calculates the ion diffusion path only by lattice constant and atomic position information, which is mainly used by considering ideal values of ions placed at atomic positions. If the ideal numerical value of the test ion, which is placed at the position x in an unit cell, and bond valence (s_j) are calculated by the distance ($d_j(x)$) from number j nearest cation with the following equation [93]

$$s_j(\mathrm{x}) = \exp[\frac{R_0 - d_j(x)}{b}] \tag{1.68}$$

Therefore, the bond-valence sum is

$$BVS(x) = \sum_{j=1}^{N_C}[s_j(x)] = \sum_{j=1}^{N_C}[m_j \exp(\frac{R_0 - d_j(x)}{b})] \tag{1.69}$$

If x is the atomic position of oxygen, BVS (x) is theoretically − 2 (unit is indicated by valence unit, vu). R_0 and b are empirical parameters, and values are set for each bond type [94]. m_j and N are the occupation ratio of ions that bonded to the test ions and the number of bonded ions, respectively. The bond valence over a certain distance has been rounded up to the calculation [95]. BV method can also be used as a method to investigate the validity of the results of a refined crystal structure. The study of the diffusion path of oxide ions by bond-valence method is carried out by using crystal structure data obtained by refinement of the diffraction pattern. Compute the BVS in the unit cell with a spatial resolution of 0.1 Å and calculate the difference between the formal charge of the oxide ion (DBVS). DBVS = 0.1 vu, the equivalent surface is yellow, whereas the equivalent surface is blue when DBVS = − 0.1 vu. A region where DBVS is close to 0 indicates that the oxide ion is a relatively stable region. It indicates that it diffuses isotropically while keeping a certain distance from the [95] cations. Pseudo-Coulomb potential of oxide-ion was calculated using BVS method [96, 97] as follows:

$$\mathrm{BVE} = D_0\left[\sum_{j=1}^{N_C}\frac{\left(s_{M-X_j} - s_{\min}\right)}{s_{\min}^2} - N_C\right] + \sum_{j=1}^{N_M} E_{\mathrm{Coulomb}}\{M - M_i\} \tag{1.70}$$

where $s_{\min}$ is the BV calculated using the bond length of the most stable energy state; D_0 is the Morse interaction potential that indicates the bond dissociation energy. In the actual calculation, empirical parameters are given for D_0 for each bond type; $s_{\min}$ are set for each bond type. E_{coulomb} is a Coulomb repulsion term between semiempirical ions.

$$\mathrm{E}_{\mathrm{Coulomb}}(M1 - M2) = \frac{q_{M1}q_{M2}}{R_{M1-M2}}\mathrm{erfc}\left(\frac{R_{M1-M2}}{\rho_{M1-M2}}\right) \tag{1.71}$$

where q_{M1} and q_{M2} are charge of $M1$ ion and $M2$ ion, respectively, $R_{\mathrm{M1-M2}}$ are $M1 - M2$ distances, erfc is a complementary error function, and $\rho_{\mathrm{M1-M2}}$ is an empirical parameter. By calculating the oxide ion energy, we can estimate the energy barrier of oxide-ion diffusion by evaluating the hopping energy barrier that contributes to the activation energy, it is possible to lead the development of ion conductors.

We selected the chemical composition as candidates for oxide-ion conductors by screening 84 Dion–Jacobson phases with the BV method [98, 99]. Because of its simple calculation procedure, the BV method is more efficient than density functional theory (DFT)-based and molecular dynamics (MD) calculations for exploring oxide-ion conductors. BV-based energy barriers E_b for oxide-ion migration were estimated by its crystallographic parameters published in the inorganic crystal structure database ICSD [100] using software softBV [96, 101]. The bond-valence parameters used in the calculations were given by Adams in the software [96, 98, 102]. The spatial resolution was set to be 0.1 Å. The E_b of Dion–Jacobson phases were estimated using BVELs and shown in Table 1.1.

Table 1.1 Bond-valence-based energy barrier for oxide-ion migration (E_b)

No	Composition	E_b (eV)	ICSD code
1	$RbLaTa_2O_7$	0.83	86,208
2	$RbLaTa_2O_7$	0.89	81,870
3	$RbNdTa_2O_7$	0.55	258,767
4	$RbLaNb_2O_7$	0.83	185,293
5	$RbLaNb_2O_7$	0.90	185,292
6	$RbCa_2Nb_3O_{10}$	0.51	260,289
7	$RbCa_2Ta_3O_{10}$	0.45	89,010
8	$RbSr_2Nb_3O_{10}$	0.79	93,674
9	$RbLa_2Ti_2TaO_{10}$	0.74	93,670
10	$RbLa_2Ti_2NbO_{10}$	0.90	90,000
11	$RbBi_2Ti_2NbO_{10}$	0.48	252,331
12	$RbPrNb_2O_7$	0.70	238,794
13	$RbNdNb_2O_7$	0.65	258,765
14	$Rb_{0.9}SrNb_2O_6F$	0.73	92,867
15	$RbNaCa_2Nb_4O_{13}$	0.67	88,853
16	$RbNa_{0.6}Sr_{0.4}Ca_2Nb_4O_{13}$	0.66	88,854
17	$RbNa_{0.86}Ca_2Nb_3O_{10}$	0.59	97,678
18	$CsCaLaTiNb_2O_{10}$	0.90	90,624
19	$CsLa_2Ti_2NbO_{10}$	0.73	90,623
20	$CsPr_2Ti_2NbO_{10}$	0.57	90,625
21	$CsNd_2Ti_2NbO_{10}$	0.46	90,626
22	$CsSm_2Ti_2NbO_{10}$	0.43	90,627
23	$CsBi_2Ti_2NbO_{10}$	0.50	252,331
24	$CsCa_2Ta_3O_{10}$	0.44	89,011

(continued)

Table 1.1 (continued)

No	Composition	E_b (eV)	ICSD code
25	$CsSrTa_2O_7$	0.72	93,677
26	$CsNdTa_2O_7$	0.52	258,766
27	$CsCa_2Nb_3O_{10}$	0.55	201,425
28	$CsSr_2Nb_3O_{10}$	0.82	93,675
29	$CsBa_2Nb_3O_{10}$	1.11	93,676
30	$CsLaNb_2O_7$	0.84	82,378
31	$CsNdNb_2O_7$	0.67	96,511
32	$CsBiNb_2O_7$	0.55	162,984
33	$CsLa_2Ti_2TaO_{10}$	0.80	93,669
34	$CsBi_2Ti_{1.93}Ta_{1.07}O_{10}$	0.53	252,332
35	$Cs_{1.09}PrNb_{2.06}O_7$	0.75	238,796
36	$KCa_2Ta_3O_{10}$	0.49	89,009
37	$KCa_2Nb_3O_{10}$	0.70	91,098
38	$KCa_2Nb_3O_{10}$	0.55	157,839
39	$KLaNb_2O_7$	0.90	72,564
40	$KLaTa_2O_7$	0.90	83,909
41	$LiLaTa_2O_7$	0.85	188,780
42	$Li_{1.8}La_2VTi_3O_{11}$	0.90	171,438
43	$LiCa_2Ta_3O_{10}$	0.46	89,006
44	$LiLaNb_2O_7$	0.83	72,566
45	$Li_{0.9}SrNb_2O_6$	0.75	92,868
46	LiLaTa2O7	1.8	81,872
47	$LiLaTa_2O_{6.14}N_{0.58}$	0.95	188,782
48	$LiLaTa_2O_{6.74}N_{0.18}$	1.01	188,781
49	$NaLaTa_2O_7$	0.90	84,363
50	$NaCa_2Ta_3O_{10}$	0.51	88,498
51	$NaCa_2Nb_3O_{10}$	0.51	97,699
52	$NaLaNb_2O_7$	0.93	72,565
53	$Na_{1.26}La_{0.913}Nb_2O_7$	0.87	67,736
54	$NaLa_2Ti_2NbO_{10}$	0.90	51,577
55	$NaPr_2Ti_2NbO_{10}$	0.60	51,578
56	$NaPr_2Ti_2NbO_{10}$	0.50	51,581
57	$NaNd_2Ti_2NbO_{10}$	0.50	51,579
58	$NaSm_2Ti_2NbO_{10}$	0.45	51,580
59	$AgLaNb_2O_7$	1.00	73,983
60	$AgLaNb_2O_7$	0.85	73,984
61	$AgLaTa_2O_7$	1.08	50,165

(continued)

Table 1.1 (continued)

No	Composition	E_b (eV)	ICSD code
62	$BaLa_2Ti_3O_{10}$	1.20	83,749
63	$BaLa_2Ti_3O_{10}$	0.82	164,868
64	$BaLa_2Ti_3O_{10}$	0.82	164,869
65	$BaLa_2Ti_3O_{10}$	0.93	164,870
66	$BaNd_2Ti_3O_{10}$	1.18	60,955
67	$BaNd_2Ti_3O_{10}$	0.57	95,516
68	$BaNd_2Ti_3O_{10}$	0.63	169,675
69	$BaNd_2Ti_3O_{10}$	0.68	194,985
70	$BaNd_2Ti_3O_{10}$	1.28	62,217
71	$BaNd_2Ti_3O_{10}$	1.18	62,218
72	$BaPr_2Ti_3O_{10}$	0.71	163,911
73	$BaPr_2Ti_3O_{10}$	0.68	163,912
74	$BaSm_2Ti_3O_{10}$	0.49	169,676
75	$HLaNb_2O_7$	0.90	72,567
76	$H_2Ca_2Nb_3O_{10.5}$	0.55	246,002
77	$H_{0.76}Rb_{2.12}S_{0.76}LaNb_2O_7$	0.90	185,290
78	$HRb_2SeLaNb_2O_7$	0.63	185,291
79	$Br_{1.05}Ca_2CuNb_3O_{10}$	0.45	163,235
80	$Br_{1.049}Ca_2Cu_{1.011}Ta_3O_{10}$	0.50	163,238
81	$Br_{0.976}Sr_2Cu_{0.959}Nb_3O_{10}$	0.66	163,237
82	$Br_{1.048}Sr_2Cu_{0.948}Ta_3O_{10}$	0.57	163,236
83	$Cl_{1.13}Cu_{0.96}PrNb_{2.08}O_7$	0.75	238,795

1.6 Maximum-Entropy Method Analysis

The maximum entropy method (MEM) is very useful to modify structural models adopted in Rietveld analysis and to determine distributions of the electron from X-ray data and distributions of nuclear densities from neutron diffraction data [103–106]. The termination effect is less serious in MEM analysis than in Fourier synthesis because MEM is capable of estimating structure factors of high reflections that have not been measured experimentally. In addition, after observed structure factors of overlapped reflections have been estimated based on the results of the Rietveld analysis, subsequent MEM analysis more or less improves structure factors, which makes MEM very effective in the extraction of the maximum amount of structural information from powder diffraction data. Three-dimensional visualization of electron and/or nuclear densities resulting from diffraction data achieves a better understanding of the highly disordered structure, chemical bonding (X-ray diffraction), or enharmonic thermal motion. For example, diffusion paths of mobile ions have been visualized in 3D from diffraction data by a sophisticated structure-refinement technique, which was called MEM-based pattern fitting (MPF). MPF is able to reveal even thermal vibrations, which are very difficult to analyze by powder diffraction due to the nonlinear least-squares method [25, 71, 104, 107–111].

There are a few variations of the MEM as applied to the Fourier inversion problem. However, these are the same in essence, there should be no significant difference in the final result for these formalisms. In this study, we follow basically Collins's formalism. The reason is that it has an explicit formulation for the actual electron density rather than the normalized density. Collins's formalism is based on the entropy expression (S)

$$S = -\sum_{r} \rho'(\mathbf{r}) \ln\left[\rho'(\mathbf{r})/\tau'(\mathbf{r})\right] \tag{1.72}$$

The probability $\rho'(\mathbf{r})$ and prior probability $\tau'(\mathbf{r})$ are connected with the actual electron density by

$$\rho'(\mathbf{r}) = \rho(\mathbf{r})/\sum_{\mathbf{r}} \rho(\mathbf{r}) \tag{1.73}$$

$$\tau'(\mathbf{r}) = \tau(\mathbf{r})/\sum_{\mathbf{r}} \tau(\mathbf{r})| \tag{1.74}$$

where $\rho(\mathbf{r})$ is the electron density at a certain pixel $\mathbf{r}$ and $\tau(\mathbf{r})$ is the prior density for $\rho(\mathbf{r})$. We introduce two types of weak constraints as given information: one is for phase-known structure factors and the other is for phase-unknown structure factors:

$$C_1 = (1/N_1) \sum_{\mathbf{k}} |F_{\text{cal}}(\mathbf{k}) - F_{\text{obs}}(\mathbf{k})|^2/\sigma^2(\mathbf{k}) \tag{1.75}$$

$$C_2 = (1/N_2) \sum_{\mathbf{k}} ||F_{\text{cal}}(\mathbf{k})| - |F_{\text{obs}}(\mathbf{k})||^2/\sigma^2(\mathbf{k}) \tag{1.76}$$

where N_1 and N_2 are the number of reflections for phase-known and -unknown structure factors, respectively; $F_{\text{obs}}(\mathbf{k})$ is the observed structure factor for reflection $\mathbf{k}$; $\sigma(\mathbf{k})$ is the standard deviation of $F_{\text{cal}}(\mathbf{k})$, which is the calculated structure factor given as

$$F_{\text{cal}}(\mathbf{k}) = V \sum_{\mathbf{r}} \rho(\mathbf{r}) \exp(-2\pi i \mathbf{r}, \mathbf{k}) \tag{1.77}$$

where V is the unit-cell volume. The expected value for C_1 and C_2 is 1. The sums are taken over the set of points for which the phase-known and -unknown structure-factor data are measured. We use Lagrange's method of undetermined multipliers to constrain C to be unity while maximizing the entropy. Then, we have

$$Q(\lambda_1, \lambda_2) = -\sum_i \rho(\mathbf{r}) \ln[\rho(\mathbf{r})/\tau(\mathbf{r})] - (\lambda_1/2)C_1 - (\lambda_2/2)C_2 \tag{1.78}$$

where λ_1 and λ_2 are Lagrange multipliers. By setting

$$\partial Q(\lambda_1, \lambda_2)/\partial\rho(\mathbf{r}) = 0 \tag{1.79}$$

and using some approximations given by Collins, the final equation for the MEM analysis is written as follows:

$$\begin{aligned}\rho(\mathbf{r}) = \exp\Big[&\ln\tau(\mathbf{r}) + (\Lambda_1/N_1)\sum\left[1/\sigma^2(\mathbf{k})\right] \times \{F_{\text{obs}}(\mathbf{k}) - F_{\text{cal}}(\mathbf{k})\}\exp(-2\pi i\mathbf{k}\cdot\mathbf{r}) \\ &+ (\Lambda_2/N_2)\sum\left[1/\sigma^2(\mathbf{k})\right] \times \{|F_{\text{obs}}(\mathbf{k})|\exp[i\varphi(\mathbf{k})] - F_{\text{cal}}(\mathbf{k})\} \\ &\times \exp(-2\pi i\mathbf{k}\cdot\mathbf{r})\Big]\end{aligned} \tag{1.80}$$

where $\Lambda_1 = \lambda_1 F_0$, $\Lambda_2 = \lambda_2 F_2$, and F_0 is equal to the number of electrons in a unit cell.

1.7 Density Functional Theory Calculation

Density functional theory (DFT) is a quantum mechanical modeling method, which is widely used in many physics and chemistry fields to study the electronic structure of many-body systems, especially in particular atoms, molecules, as well as condensed phases. An analytical solution of the many-electron Schrdinger equation is not available, and a numerical solution, while perfectly possible in theory, is effectively impossible in practice for more than a handful of electrons due to the finite speed and memory of computers. DFT-based electronic structure calculations have been widely used, and have successfully described various properties of many condensed matter systems.

In the present work, DFT calculations were used to investigate the validity of refined crystal structure, band structure, and the occupancy of Ti and Nb elements in the Dion–Jacobson phases. Crystallographic parameters, which were obtained from the Rietveld refinement of neutron diffraction data, were used as the initial model in the structural optimization. The generalized gradient approximation electronic calculations were carried out with the Vienna Ab initio Simulation Package, in order to study the optimized structure of supercells. Projector augmented-wave potentials were used for all the atoms and a plane-wave basis set with a cutoff of 500 eV was used. The Perdew – Burke – Ernzerhof GGA was used for the exchange and correlation function. Unit-cell parameters and atomic coordinates were optimized with the convergence condition.

References

1. Yang Z et al (2011) Electrochemical energy storage for green grid. Chem Rev 111(5):3577–3613
2. Mai L, Tian X, Xu X, Chang L, Xu L (2014) Nanowire electrodes for electrochemical energy storage devices. Chem Rev 114(23):11828–11862
3. Lewis NS (2015) Introduction: solar energy conversion. Chem Rev 115(23):12631–12632
4. Song C (2006) Global challenges and strategies for control, conversion and utilization of CO_2 for sustainable development involving energy, catalysis, adsorption and chemical processing. Catal Today 115(1–4):2–32
5. Feng K, Davis SJ, Sun L, Hubacek K (2015) Drivers of the US CO_2 emissions 1997–2013. Nat Commun 6:1–8
6. Zhu H et al (2016) Wood-derived materials for green electronics, biological devices, and energy applications. Chem Rev 116(16):9305–9374
7. Joselin Herbert GM, Iniyan S, Sreevalsan E, Rajapandian S (2007) A review of wind energy technologies. Renew Sustain Energy Rev 11(6):1117–1145
8. Mahato N, Banerjee A, Gupta A, Omar S, Balani K (2015) Progress in material selection for solid oxide fuel cell technology: a review. Prog Mater Sci 72:141–337
9. Steele BCH, Heinzel A (2001) Materials for fuel-cell technologies, 414(November):345–352
10. Tie SF, Tan CW (2013) A review of energy sources and energy management system in electric vehicles. Renew Sustain Energy Rev 20:82–102
11. Wang W, Su C, Wu Y, Ran R, Shao Z (2013) Progress in solid oxide fuel cells with nickel-based anodes operating on methane and related fuels. Chem Rev 113(10):8104–8151
12. Chu S, Majumdar A (2012) Opportunities and challenges for a sustainable energy future. Nature 488(7411):294–303
13. Whittingham MS, Savinell RF, Zawodzinski T (2004) Introduction: batteries and fuel cells. Chem Rev 104(10):4243–4244
14. Winter M, Brodd RJ (2004) What are batteries, fuel cells, and supercapacitors? Chem Rev 104(10):4245–4269
15. Gür TM (2013) Critical review of carbon conversion in 'carbon fuel cells.' Chem Rev 113(8):6179–6206
16. Singh B, Ghosh S, Aich S, Roy B (2017) Low temperature solid oxide electrolytes (LT-SOE): a review. J Power Sourc 339:103–135
17. Park S, Vohs JM, Gorte RJ (2000) Direct oxidation of hydrocarbons in a solid-oxide fuel cell. Nature 404(6775):265–267
18. Boldrin P, Ruiz-Trejo E, Mermelstein J, Bermúdez Menéndez JM, Ramírez Reina T, Brandon NP (2016) Strategies for carbon and sulfur tolerant solid oxide fuel cell materials, incorporating lessons from heterogeneous catalysis. Chem Rev 116(22):13633–13684
19. Minh NQ (2004) Solid oxide fuel cell technology—features and applications. Solid State Ionics 174(1–4):271–277
20. Ishihara T, Honda M, Shibayama T, Furutani H, Takita Y (1998) An intermediate temperature solid oxide fuel cell utilizing superior oxide ion conducting electrolyte, doubly doped $LaGaO_3$ perovskite. Ionics (Kiel) 4(5–6):395–402
21. Chen D, Ran R, Zhang K, Wang J, Shao Z (2009) Intermediate-temperature electrochemical performance of a polycrystalline $PrBaCo_2O_{5+\delta}$ cathode on samarium-doped ceria electrolyte. J Power Sourc 188(1):96–105

22. Wachsman ED, Lee KT (2011) Lowering the temperature of solid oxide fuel cells. Science (80–), 334(6058):935–939
23. Tarancón A (2009) Strategies for lowering solid oxide fuel cells operating temperature. Energies 2(4):1130–1150
24. Willard DE, Oster CV, Ritchie IW, Jones WW, Zogorski JS (1984) The environmental aspect of developing Indiana oil shale. Stud Environ Sci 25(C):101–118
25. Yashima M (2015) Invited review: some recent developments in the atomic-scale characterization of structural and transport properties of ceria-based catalysts and ionic conductors. Catal Today 253:3–19
26. Fop S (2016) Novel oxide-ion conductors in the hexagonal perovskite family, PhD Thesis. University of Aberdeen
27. Nernst W (1899) On the electrolytic conduction of solid bodies at high temperatures. Z Electrochem 6(2):41–43
28. Goodenough JB (2000) Oxide-ion conductors by design. Nature 404(6780):821–823
29. Boivin JC, Mairesse G (1998) Recent material developments in fast oxide-ion conductors. Chem Mater 10(10):2870–2888
30. Huang K, Wan J, Goodenough JB (2001) Oxide-ion conducting ceramics for solid oxide fuel cells. J Mater Sci 36(5):1093–1098
31. Skinner SJ, Kilner JA (2003) Oxygen ion conductors. Mater Today 3(6):30–37
32. Yashima M (2008) Crystal structures, structural disorders and diffusion paths of ionic conductors from diffraction experiments. Solid State Ionics 179(21–26):797–803
33. Thangadurai V, Weppner W (2006) Recent progress in solid oxide and lithium ion conducting electrolytes research. Ionics (Kiel) 12(1):81–92
34. Goodenough JB (2000) Oxide-ion conductors by design. Nature 404(April)
35. Singhal SC (2000) Advances in solid oxide fuel cell technology. Solid State Ionics 135(1–4):305–313
36. Kwon OH, Choi GM (2006) Electrical conductivity of thick film YSZ. Solid State Ionics 177(35–36):3057–3062
37. Lacorre P, Goutenoire Ë, Bohnke O, Retoux R, Laligant Y (2000) Designing fast oxide-ion conductors. Nature 404(April):9–11
38. Kubicek M, Bork AH, Rupp JLM (2017) Perovskite oxides-a review on a versatile material class for solar-to-fuel conversion processes. J Mater Chem A 5(24):11983–12000
39. Suntivich J, May KJ, Gasteiger HA, Goodenough JB, Shao-Horn Y (2011) A perovskite oxide optimized for oxygen evolution catalysis from molecular orbital principles. Science (80–), 334(6061):1383–1385
40. De Souza RA (2015) Oxygen diffusion in $SrTiO_3$ and related perovskite oxides. Adv Funct Mater 25(40):6326–6342
41. Wang W, Liang Zhu ZS, Ran R, Tadé M (2016) Perovskite materials in energy storage and conversion. Asia-Pacific J Chem Eng 11:338–369
42. Saiful Islam M (2000) Ionic transport in ABO3 perovskite oxides: a computer modelling tour. J Mater Chem 10(4):1027–1038
43. Ishihara T, Shibayama T, Nishiguchi H, Takita Y (2001) Oxide ion conductivity in La0.8Sr0.2Ga0.8Mg0.2-xNix O3 perovskite oxide and application for the electrolyte of solid oxide fuel cells. J Mater Sci 36(5):1125–1131
44. Ishihara T (2006) Development of new fast oxide-ion conductor and application for intermediate temperature solid oxide fuel cells. Bull Chem Soc Jpn 79(8):1155–1166
45. Li M et al (2014) A family of oxide-ion conductors based on the ferroelectric perovskite Na0.5Bi0.5TiO_3. Nat Mater 13(1):31–35
46. Roukos R, Zaiter N, Chaumont D (2018) Relaxor behaviour and phase transition of perovskite ferroelectrics-type complex oxides (1–x)Na0.5Bi0.5TiO_3–x$CaTiO_3$ system, 7(2):124–142
47. Yang F, Li M, Li L, Wu P, Pradal-Velázquez E, Sinclair D (2017) Optimisation of oxide-ion conductivity in acceptor-doped $Na_{0.5}Bi_{0.5}TiO_3$ perovskite: approaching the limit? J Mater Chem A 21658–21662

48. Goodenough JB, Ruiz-Diaz JE, Zhen YS (1990) Oxide-ion conduction in $Ba_2In_2O_5$ and $Ba_3In_2MO_8$ (M=Ce, Hf, or Zr). Solid State Ionics 44(1–2):21–31
49. Jeen H et al (2013) Reversible redox reactions in an epitaxially stabilized SrCoO x oxygen sponge. Nat Mater 12(11):1057–1063
50. Kagomiya I et al (2017) Weak ferromagnetic ordering in brownmillerite $Ca_2Fe_2O_5$ and its effect on electric field gradients. Phys Chem Chem Phys 19(46):31194–31201
51. Amow G, Skinner SJ (2006) Recent developments in Ruddlesden-Popper nickelate systems for solid oxide fuel cell cathodes. J Solid State Electrochem 10(8):538–546
52. Yamada A et al (2008) Ruddlesden-popper-type epitaxial film as oxygen electrode for solid-oxide fuel cells. Adv Mater 20(21):4124–4128
53. Wang TH, Henderson CN, Draskovic TI, Mallouk TE (2014) Synthesis, exfoliation, and electronic/protonic conductivity of the Dion-Jacobson phase layer perovskite $HLa_2TiTa_2O_{10}$. Chem Mater 26(2):898–906
54. Ruddlesden SN, Popper P (1958) The compound $Sr_3Ti_2O_7$ and its structure. Acta Crystallogr 11(1):54–55
55. Patrakeev MV, Leonidov IA, Kozhevnikov VL, Kharton VV (2004) Ion-electron transport in strontium ferrites: relationships with structural features and stability. Solid State Sci 6(9):907–913
56. Fujii K et al (2014) New perovskite-related structure family of oxide-ion conducting materials $NdBaInO_4$. Chem Mater 26(8):2488–2491
57. Fujii K, Shiraiwa M, Esaki Y, Yashima M, Kim SJ, Lee S (2015) Improved oxide-ion conductivity of $NdBaInO_4$ by Sr doping. J Mater Chem A 3(22):11985–11990
58. Yang X, Liu S, Lu F, Xu J, Kuang X (2016) Acceptor doping and oxygen vacancy migration in layered perovskite $NdBaInO_4$-based mixed conductors. J Phys Chem C 120(12):6416–6426
59. Ishihara T, Yan Y, Sakai T, Ida S (2016) Oxide ion conductivity in doped $NdBaInO_4$. Solid State Ionics 288:262–265
60. Fop S et al (2016) Oxide ion conductivity in the hexagonal perovskite derivative $Ba_3MoNbO_{8.5}$. J Am Chem Soc 138(51):16764–16769
61. Fop S, McCombie KS, Wildman EJ, Skakle JMS, Mclaughlin AC (2019) Hexagonal perovskite derivatives: a new direction in the design of oxide ion conducting materials. Chem Commun 55(15):2127–2137
62. Yashima M et al (2019) Direct evidence for two-dimensional oxide-ion diffusion in the hexagonal perovskite-related oxide $Ba_3MoNbO_{8.5-\delta}$. J Mater Chem A 7(23):13910–13916
63. Iharada T, Hammouche A, Fouletier J, Kleitz M, Boivin JC, Mairesse G (1991) Electrochemical characterization of BIMEVOX oxide-ion conductors. Solid State Ionics 48(3–4):257–265
64. Pirovano C, Steil MC, Capoen E, Nowogrocki G, Vannier RN (2005) Impedance study of the microstructure dependence of the electrical properties of BIMEVOXes. Solid State Ionics 176(25–28):2079–2083
65. Piva RH, Biz H, Piva DH, Morelli MR (2016) Facile preparation of BIMEVOX powders via melting process: from synthesis to sintering optimization. Ceram Int 42(6):7088–7098
66. Abraham F, Boivin JC, Mairesse G, Nowogrocki G (1990) The bimevox series: A new family of high performances oxide-ion conductors. Solid State Ionics 40–41(PART 2):934–937
67. Goff RJ, Keeble D, Thomas PA, Ritter C, Morrison FD, Lightfoot P (2009) Leakage and proton conductivity in the predicted ferroelectric $CsBiNb_2O_7$. Chem Mater 21(7):1296–1302
68. Strayer ME et al (2016) Emergent Noncentrosymmetry and piezoelectricity driven by oxygen octahedral rotations in n = 2 Dion-Jacobson phase layer perovskites. Adv Funct Mater 26(12):1930–1937
69. Collins C et al (2017) Accelerated discovery of two crystal structure types in a complex inorganic phase field. Nature 546(7657):280–284
70. Yashima M, Ali R (2009) Structural phase transition and octahedral tilting in the calcium titanate perovskite $CaTiO_3$. Solid State Ionics 180(2–3):120–126
71. Yashima M, Kamioka T (2008) Neutron diffraction study of the perovskite-type lanthanum cobaltite. Solid State Ionics 178:1939–1943

72. Yashima M, Tanaka M (2004) Performance of a new furnace for high-resolution synchrotron powder diffraction up to 1900 K: application to determine electron density distribution of the cubic $CaTiO_3$ perovskite at 1674 K. J Appl Crystallogr 37(5):786–790
73. Toraya H, Hibino H, Ohsumi K (1996) A new powder diffractometer for synchrotron radiation with a multiple-detector system. J Synchrotron Radiat 3(Pt 2):75–83
74. Yashima M, Oh-uchi K, Tanaka M, Ida T (2006) A compact furnace for synchrotron powder diffraction experiments up to 1800 K. J Am Ceram Soc 89(4):1395–1399
75. Kawaguchi S et al (2017) High-throughput powder diffraction measurement system consisting of multiple MYTHEN detectors at beamline $BL0_2B_2$ of SPring-8. Rev Sci Instrum 88(8)
76. Nishibori E et al (2001) The large Debye-Scherrer camera installed at SPring-8 $BL0_2B_2$ for charge density studies. J Phys Chem Solids 62(12):2095–2098
77. Young RA (1993) The Rietveld method. Oxford University Press, New York
78. Kawamura K et al (2015) Structural origin of the anisotropic and isotropic thermal expansion of K_2NiF_4-type $LaSrAlO_4$ and Sr_2TiO_4. Inorg Chem 54(8):3896–3904
79. Yonemura M et al (2014) Development of SPICA, new dedicated neutron powder diffractometer for battery studies. J Phys Conf Ser 502(1):0–4
80. Ishigaki T et al (2009) IBARAKI materials design diffractometer (iMATERIA)-Versatile neutron diffractometer at J-PARC. In: Nuclear instruments and methods in physics research section A accelerators spectrometers detectors and associated equipment abbreviation, vol 600, no 1, pp 189–191
81. Von Dreele RB, Jorgensen JD, Windsor CG (1982) Rietveld refinement with spallation neutron powder diffraction data. J Appl Crystallogr 15(6):581–589
82. Torii S et al (2011) Super high resolution powder diffractometer at J-PARC. J Phys Soc Jpn 80(Suppl B):3–6
83. Oishi R et al (2009) Rietveld analysis software for J-PARC. In: Nuclear instruments and methods in physics research section A accelerators spectrometers detectors and associated equipment abbreviation, vol 600, no 1, pp 94–96
84. Kajitani M et al (2005) In situ neutron diffraction study on fast oxide-ion conductor $LaGaO_3$-based perovskite compounds. Chem Mater 17(16):4235–4243
85. Slater PR, Irvine JTS, Ishihara T, Takita Y (1998) High-temperature powder neutron diffraction study of the oxide-ion conductor La0.9Sr0.1Ga0.8Mg0.2O2.85. J Solid State Chem 143(139):135–143
86. Yashima M, Sirikanda N, Ishihara T (2010) Crystal structure, diffusion path, and oxygen permeability of a Pr_2NiO_4-based mixed conductor $(Pr_{0.9}La_{0.1})_2(Ni_{0.74}Cu_{0.21}Ga_{0.05})O_{4+\delta}$. J Am Chem Soc 132(7):2385–2392
87. Young RA, Wiles DB (1982) Profile shape functions in Rietveld refinements. J Appl Crystallogr 15(4):430–438
88. Von Dreele RB (1997) Quantitative texture analysis by Rietveld refinement. J Appl Crystallogr 30(4):517–525
89. Mccusker LB, Von Dreele RB, Cox DE, Louër D, Scardi P (1999) Rietveld refinement guidelines. J Appl Crystallogr 32(1):36–50
90. Kröger FA (1964) The Chemistry of Imperfect Crystals, North-Holland, Amsterdam, and Wiley, New York.
91. West A, Irvine J, Sinclair D (1990) Electroceramics: characterization by impedance spectroscopy. Adv Mater 2(3):132–138
92. Shih DPC, Aguadero A, Skinner SJ (2018) Improvement of ionic conductivity in A-site lithium doped sodium bismuth titanate. Solid State Ionics 317(January):32–38
93. Preiser C, Lösel J, Brown ID, Kunz M, Skowron A (1999) Long-range Coulomb forces and localized bonds. Acta Crystallogr Sect B Struct Sci 55(5):698–711
94. http://www.iucr.org/resources/data/datasets/bond-valence-parameters
95. Brown ID, Poeppelmeier KR (Eds.) (2014) Bond valences. Berlin: Springer Berlin Heidelberg.
96. Adams S (2006) From bond valence maps to energy landscapes for mobile ions in ion-conducting solids. Solid State Ionics 177(19–25):SPEC. ISS 1625–1630

97. Janek J, Martin M, Becker KD (2009) Physical chemistry of solids--the science behind materials engineering. Phys Chem Chem Phys 11(17):3010
98. Adams S (2001) Relationship between bond valence and bond softness of alkali halides and chalcogenides. Acta Crystallogr Sect B Struct Sci 57(3):278–287
99. Avdeev M, Sale M, Adams S, Rao RP (2012) Screening of the alkali-metal ion containing materials from the Inorganic Crystal Structure Database (ICSD) for high ionic conductivity pathways using the bond valence method. Solid State Ionics 225:43–46
100. Belkly A, Helderman M, Karen VL, Ulkch P (2002) New developments in the Inorganic Crystal Structure Database (ICSD): accessibility in support of materials research and design. Acta Crystallogr Sect B Struct Sci 58(3), Part 1:364–369
101. Chen H, Wong LL, Adams S (2019) SoftBV—a software tool for screening the materials genome of inorganic fast ion conductors. Acta Crystallogr Sect B Struct Sci Cryst Eng Mater 75(1):18–33
102. Adams S (2000) Modelling ion conduction pathways by bond valence pseudopotential maps. Solid State Ionics 136–137:1351–1361
103. Sakata M, Sato M (1990) Accurate structure analysis by the maximum-entropy method. Acta Crystallogr Sect A 46(4):263–270
104. Sakata M, Uno T, Takata M, Howard CJ (1993) Maximum-entropy-method analysis of neutron diffraction data. J Appl Crystallogr 26(pt 2):159–165
105. Takata M et al (1995) Confirmation by X-ray diffraction of the endohedral nature of the metallofullerene Y@C82. Nature 377(6544):46–49
106. Tanaka M, Yashima M, Mori T (2004) A new furnace for high-temperature synchrotron X-ray powder diffraction experiments—electron density analysis by powder X-ray diffraction at 1300 °C. AIP Conf Proc 705(June):1055–1058
107. Ali R, Yashima M, Izumi F (2007) Diffusion path of oxide ions in an oxide-ion conductor $La_{0.64}(Ti_{0.92}Nb_{0.08})O_{2.99}$ with a double perovskite-type structure. Chem Mater 19(13):3260–3264
108. Yashima M, Kobayashi S, Yasui T (2007) Positional disorder and diffusion path of oxide ions in the Yttria-doped ceria $Ce_{0.93}Y_{0.07}O_{1.96}$. Faraday Discuss 134:369–376
109. Bhosale DR, Yusuf SM, Kumar A, Mukadam MD, Patil SI (2017) High oxide ion conductivity below 500 C in the garnets L ax Y3–x F e5 O12+δ. Phys Rev Mater 1(1):2–6
110. Yashima M et al (2019) Direct evidence for two-dimensional oxide-ion diffusion in the hexagonal perovskite-related oxide $Ba_3MoNb_{O8.5}$–δ. J Mater Chem A 13910–13916
111. Yashima M, Precise structure analysis of inorganic materials for clean energy by maximum-entropy method: neutron and synchrotron X-Ray powder diffraction studies, 4:978–981

Chapter 2
New Oxide-Ion Conductors of Dion–Jacobson-Type Layered Perovskites $CsBi_2Ti_2NbO_{10-\delta}$

Abstract Oxide-ion conductors exhibit many important applications in many electrochemical devices, for example, solid-oxide fuel cells (SOFCs), solid-oxide electrolysis cell (SOECs), oxygen gas sensors, and oxygen separation membranes. Dion–Jacobson phases have attracted attention because of their wide chemical and physical performances, however, up to now, there are no reports of oxide-ion conductors with the Dion–Jacobson structure. Here, we have reported the new oxide-ion conductors of Dion–Jacobson-type layered perovskites for the first time. The oxide-ion migration energy barriers, E_b, of 69 Dion–Jacobson phases were calculated using 83 Dion–Jacobson-type crystal data, in order to find new oxide-ion conductors in the Dion–Jacobson phases. $CsBi_2Ti_2NbO_{10-\delta}$ was chosen because of its relatively low E_b. The bulk conductivity of $CsBi_2Ti_2NbO_{10-\delta}$ (δ is the oxygen vacancy content) was 8.9×10^{-2} S cm^{-1} at 1073 K, which was higher than that of the conventional yttria-stabilized zirconia (YSZ). There is a jump in conductivity between 823 and 873 K, which may can be attributed to the high level of oxygen vacancy concentration and the orthorhombic-to-tetragonal phase transition. In order to improve its conductivity, doping elements in the *A* (Bi site) and *B* sites (Ti/Nb site) were investigated. The $CsBi_{2-x}M_xTi_{2+y}Nb_{1-y}O_{10-x/2-y/2-\delta}$ (M = alkaline earth metal; x = 0.1, 0.2, 0.3, 0.4, 0.5, and 0.6; y = 0.2, 0.1, −0.1, and −0.2) compounds were synthesized and the electrical properties were investigated. The high conductivity of Dion–Jacobson-type layered perovskites $CsBi_2Ti_2NbO_{10-\delta}$ can be attributable to the large anisotropic thermal motions of oxygen atoms (especially O1 and O2), the existence of carrier (oxygen vacancy in high-temperature ranges) and the formation of an oxide-ion conducing layer with large bottleneck sizes (inner perovskite layers). The present finding of high oxide-ion conductivities in the Dion–Jacobson-type layered perovskites $CsBi_2Ti_2NbO_{10-\delta}$ would open new avenues in the design and discovery of other novel oxide-ion conductors. Parts of this paper have been published in the following journal articles. Zhang, W., Fujii, K., Niwa, E., Hagihala, M., Kamiyama, T., & Yashima, M. (2020). Oxide-ion conduction in the Dion–Jacobson phase $CsBi_2Ti_2NbO_{10-\delta}$.*Nature communications*, *11*(1), 1–8.

W. Zhang, *Oxide-Ion Conduction in the Dion–Jacobson-Type Layered Perovskites*, Springer Theses, https://doi.org/10.1007/978-981-19-2247-3_2

2.1 Introduction

Oxide-ion (O^{2-}) conductors, also called oxygen ion conductors, have attracted much attention because of their extensive applications such as solid oxide electrolysis cells (SOECs), solid-oxide fuel cells (SOFCs), gas sensors, oxygen separation membranes, and catalysts [1–10]. However, because of the oxide ions and the cation network interactions, novel oxide-ion conductors have been achieved in a limited number of crystal structure families, such as perovskite, fluorite, melilite, and apatite structures [11–17]. As the limited crystal families restrict their further development and application, the discovery of new and novel oxide-ion conductors with a new crystal structure is very important. Many layered perovskites such as BIMEVOX, [18–21] Aurivillius phase, [22–24] Ruddlesden–Popper phase, [25–29] double perovskite, [30–32] brownmillerite, [33, 34] hexagonal perovskite derivatives, [35–37] and $BaNdInO_4$-based oxides [38–40] have been reported to show high ionic conductivities. Also as a layered perovskite phase, the Dion–Jacobson phase is an *A'*/*A* cation-ordered layered perovskite with a chemical formula $A'\ [A_{n-1}B_nO_{3n+1}]$ [41, 42]. Similar to other layered perovskite phases, thus, oxide-ion conduction can also be expected in the Dion–Jacobson phases. There have been numerous studies on the electrical properties such as ferroelectricity, [43] proton conduction, [44, 45] lithium-ionic conduction, [46, 47] sodium-ionic conduction, [47, 48] and mixed proton–electron conduction [49] in the Dion–Jacobson phases. However, there have been no reports about oxide-ion conduction in the Dion–Jacobson-type layered perovskites. In the present study, we report oxide-ion conduction in the Dion–Jacobson-type layered perovskites for the first time.

Here, we have screened about 70 Dion–Jacobson phases using the bond-valence method (Figs. 2.1 and 2.2) [12, 16, 50–55]. The chemical composition $CsBi_2Ti_2NbO_{10}$ was selected, because the E_b was as low as 0.5 eV. What is more, there were no rare-earth elements in the compound $CsBi_2Ti_2NbO_{10}$. We found that the bulk ionic conductivities σ_b of Dion–Jacobson phase $CsBi_2Ti_2NbO_{10-\delta}$ were 8.9×10^{-2} S cm^{-1} at 1073 K, which were higher than those of the conventional YSZ in the high-temperature ranges. The present discovery of the new crystal structure family, Dion–Jacobson-type layered perovskite family, of oxide-ion conductors would provide new possibilities in the discovery of novel oxide-ion conductors, owing to the high oxide-ion conductivities of $CsBi_2Ti_2NbO_{10-\delta}$ and the wide range of chemical compositions in the Dion–Jacobson phases.

2.2 Experimental Section

2.2.1 Synthesis and Characterization

$CsBi_2Ti_2NbO_{10-\delta}$ and $CsBi_{2-x}M_xTi_{2+y}Nb_{1-y}O_{10-x/2-y/2-\delta}$ (M = alkaline-earth metal; x = 0.1, 0.2, 0.3, 0.4, 0.5 and 0.6; y = 0.2, 0.1, −0.1, and −0.2) compounds were

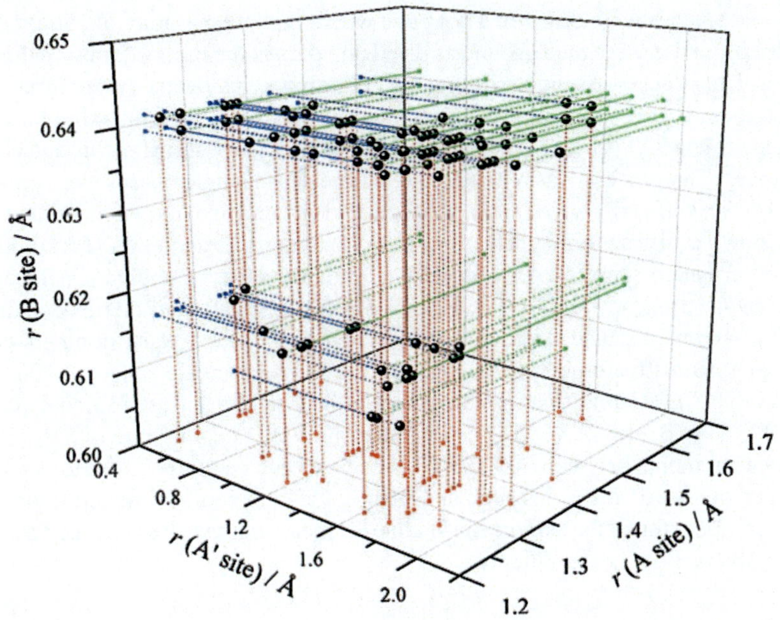

Fig. 2.1 Structure field map of Dion–Jacobson-type layered perovskite structure with a formula of $A''\ [A_{n-1}B_nO_{3n+1}]$

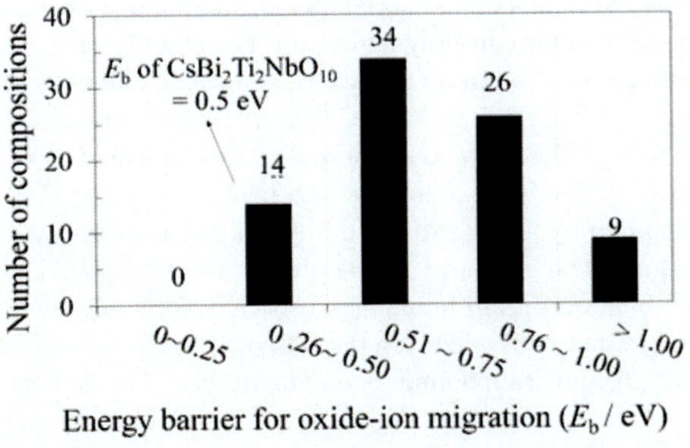

Fig. 2.2 Histogram of the bond-valence-based energy barriers E_b for the oxide-ion migration of the 83 crystal data

synthesized using the solid-state-reaction method. Starting materials of Cs_2CO_3, Bi_2O_3, TiO_2, Nb_2O_5, and alkaline earth oxides with higher purity than 99.9% were mixed in an agate mortar as ethanol slurries for more than half an hour. 1.3 mol% excess amount of Cs_2CO_3 was used for all the compounds to compensate for the mass loss because of the volatilization of the cesium species when sintering [43]. and then, the mixed slurries were dried on a hot plate with a high temperature of about 100 °C. The dried slurries were then ground as dried powders in the mortar for more than half an hour. This process of mixing, drying, and grinding was repeated two or three times. At last, the mixtures were pressed into pellets at about 100 MPa uniaxially and then sintered in air at 900 °C. for 12 h with the rate of 5 °C min^{-1} during heating and cooling progresses. The process was repeated a few times until single phases were obtained. Parts of the sintered pellets were crushed and ground into powders to carry out X-ray powder diffraction, synchrotron X-ray powder diffraction (SXPRD), atomic absorption spectroscopy (AAS, Hitachi Z-2300), inductively coupled plasma optical emission spectroscopy (ICP-OES, Hitachi PS3520UVDD), and thermogravimetric (TG) measurements. The existing phase of $CsBi_2Ti_2NbO_{10-\delta}$ was examined with an X-ray powder diffractometer (RINT-2550, Cu $K\alpha$ radiation, 2θ range: 5–100°, Step interval: 0.02°). TG analysis was carried out in flowing dry and static air between RT and 800 °C by Bruker-AXS 2020SA instrument with a heating rate of 1 °C min^{-1}. The measurements between 200 and 800 °C were repeated three times to negate the influence of absorbed species such as water and also to confirm the reproducibility. The decomposition of the sample during the TG measurements was checked by X-ray powder diffraction measurements.

2.2.2 Electrical Conductivity Measurements

The electrical conductivities of $CsBi_2Ti_2NbO_{10-\delta}$ compounds were measured using AC impedance spectroscopy in flowing dry air, N_2, and O_2 gases. A sintered pellet with 20 mm in diameter, 2.6 mm in thickness, and relative density of 91% was used for this measurement from 300 to 900 °C. The Au paste was used as electrodes on all sides of the pellet. The surface of the pellet was polished in order to decrease resistance, such as electrode resistance and exchange resistance. The AC impedance spectra were recorded by a Solartron 1260 impedance analyzer, with the measured condition of frequency range from 15 MHz to 1 Hz and an applied alternating voltage of 100 mV. The bulk and grain boundary responses were observed and evidenced at high frequencies and equivalent circuit analyses were carried out to extract the bulk conductivity σ_b and grain boundary conductivity σ_{gb} at each temperature range using the ZView. Temperature dependence of total DC electrical conductivities σ_{tot} of $CsBi_2Ti_2NbO_{10-\delta}$ and $CsBi_{2-x}M_xTi_{2+y}Nb_{1-y}O_{10-x/2-y/2-\delta}$ (M = alkaline-earth metal; $x = 0.1, 0.2, 0.3, 0.4, 0.5.$ and 0.6; $y = 0.2, 0.1, -0.1$, and -0.2) were also measured in dry and wet air on cooling by a DC 4-probe method, with pellets about 4–5 mm in diameter, 10–15 mm in height and relative density of 80–90%. The oxygen partial pressure $P(O_2)$ dependence of total DC electrical conductivities were measured by the DC 4-probe method at different temperatures in the $P(O_2)$ region about between 3×10^{-23} and 1 atm using mixture gases of O_2, N_2, and 5%H_2 in N_2.

2.2.3 Neutron and Synchrotron X-ray Diffraction Measurements

High-angular-resolution SXRPD data of $CsBi_2Ti_2NbO_{10-\delta}$ were measured in air between 297 and 1073 K using six one-dimensional solid-state detectors at the BL02B2 beam line of SPring-8, Japan [56, 57]. The $CsBi_2Ti_2NbO_{10-\delta}$ powder was loaded into a quartz glass capillary for the SXRPD measurements, where the inner diameter of the glass capillary is 0.1 or 0.2 mm. The wavelength for the SXRPD measurements was 0.6992620(4) Å using silicon powder (NIST SRM 640c). The absorption corrections were performed using the empirical formula reported by Rouse and Cooper [58].

High-temperature neutron-diffraction measurements were measured using the super-high-resolution time-of-flight (TOF) neutron diffractometer SuperHRPD installed at the Materials and Life Science Experimental Facility of J-PARC, Japan [59, 60]. The measurements were carried out in vacuum from RT to high temperatures. The absorption correction was also performed using the same method in the SXRPD measurements [61]. All of the diffraction data were analyzed by the Rietveld method using the Z-Rietveld program [62]. Oxygen contents in the $CsBi_2Ti_2NbO_{10-\delta}$ compound were calculated using the oxygen occupancy factors refined using the neutron-diffraction data and TG analysis. The oxygen contents in $CsBi_2Ti_2NbO_{10-\delta}$ were calculated to be 9.834(6) at 600 °C, 9.823(6) at 700 °C, and 9.802(6) at 800 °C using the neutron-diffraction data, which agreed well with those calculated from the TG data: 9.839 at 600 °C, 9.828 at 700 °C, and 9.807 at 800 °C.

2.2.4 Bond-Valence-Based Energy Calculation

The BVELs of $CsBi_2Ti_2NbO_{10-\delta}$ were calculated using the refined crystal structure, and then to examine the oxide-ion diffusion pathway in the unit cell. The spatial resolution in the calculations was also set to 0.1 Å. The BVELs were also used to investigate possible ion migration pathways in the structure and E_b for all the ion migration, including oxide ins, Cs ions, and other cations. At last, the calculated BVELs with their structures were drawn with the software of VESTA [63].

2.2.5 Maximum-Entropy Method Analysis

The neutron scattering length density distribution was investigated using the maximum-entropy method (MEM). MEM method was performed by the Dysnomia [64] using the structure factors obtained from the Rietveld refinement with the neutron-diffraction data at high temperatures. The MEM method was carried out with a unit cell, which was divided into 38 × 38 × 156 pixels for calculations. The structure, neutron scattering length density distributions were depicted with the software of VESTA [63].

2.2.6 Oxygen Concentration-Cell Measurements

Oxygen concentration-cell measurements were carried out to calculate its oxide-ion transport number t_{ion} in the $CsBi_2Ti_2NbO_{10\text{-}\delta}$, where use a sintered pellet with 20 mm in diameter, 4.5 mm in height, and a relative density of 91%. The synthesized pellet was attached to an alumina tube with a glass seal. One side of the pellet was exposed to a flowing dry air and the other side to a flowing dry O_2 (Air/O_2), N_2 (Air/N_2), and 5%H_2 in N_2 (Air/5%H_2 in N_2) gases between 600 and 900 °C. A Keithley model 617 electrometer was used to measure the electromotive forces of the oxygen concentration cell. The following Nernst equation was used to estimate the t_{ion}, where F is the Faraday constant, R is the gas constant, and T is the absolute temperature, $P(O_2)$ is the oxygen partial pressure of the oxygen and nitrogen gas mixture, and $P^0(O_2)$ (=0.21 atm) is the oxygen partial pressure of the dry air.

$$E = t_{ion}\frac{RT}{4F}\ln\left(\frac{P(O_2)}{P^0(O_2)}\right)$$

The activation energies E_a for the conductivities were estimated using the Arrhenius equation:

$$\sigma = \frac{A_0}{T}\exp\left(-\frac{E_a}{kT}\right)$$

where A_0, k, and T are the pre-exponential factor, Boltzmann constant, and absolute temperature, respectively.

The pellet after impedance spectroscopy and oxygen concentration-cell measurements was crushed and ground into powder. The X-ray diffraction measurements were carried out using the ground powders, to check the phase stability of the $CsBi_2Ti_2NbO_{10\text{-}\delta}$ at different temperature ranges and different $P(O_2)$ atmospheres.

2.2.7 Improved Oxide-Ion Conductivity by Doping

Generally, doping with different cations is undesirable for the ionic conductor, because it can lead to the appearance of electron conductivities or hole conductivities. On the other hand, it was also reported that the oxide-ion conductivities can be improved by doping, which increase the concentration of oxygen vacancies and or interstitial oxygen by exchanging smaller cations with the larger cation. Sometimes, the doping may also lower the activation energy, which improves the conductivities as low-temperature ranges. In the present work, $CsBi_{2-x}M_xTi_{2+y}Nb_{1-y}O_{10-x/2-y/2-\delta}$ (M = alkaline-earth metal; $x = 0.1, 0.2, 0.3, 0.4, 0.5$, and 0.6; $y = 0.2, 0.1, -0.1$, and -0.2) compounds were synthesized and the electrical properties were investigated.

2.3 Results and Discussion

We selected the chemical composition $CsBi_2Ti_2NbO_{10-\delta}$ as a candidate for oxide-ion conductors by screening about 70 Dion–Jacobson phases through the BV method [53–55]. Because of its simple calculation procedure, the BV method is more efficient than DFT and MD calculations for exploring oxide-ion conductors. The BVELs for a test oxide ion in 80 compounds with the Dion–Jacobson structure were examined to search for oxide-ion conductors. The BVE was calculated using its crystallographic parameters from the ICSD [65] with the software of softBV [66, 67]. The bond-valence parameters used in the calculations were given by Adams in the software [53, 66, 60]. The E_b of $CsBi_2Ti_2NbO_{10}$ was relatively low (E_b = 0.5 eV). Furthermore, the BV-based energy of $CsBi_2Ti_2NbO_{10-\delta}$ was calculated with the crystal parameters refined using in situ high-temperature neutron-diffraction data taken at 700 °C in order to study the anisotropic thermal motions and the oxide-ion diffusion pathways in the composition $CsBi_2Ti_2NbO_{10-\delta}$.

Rietveld refinements of the SXRD data taken in air and neutron diffraction data taken in vacuum of $CsBi_2Ti_2NbO_{10-\delta}$ were carried out by a single orthorhombic phase (space group of *Ima*2) from RT to 813 K and a single tetragonal phase (space group of *P*4/*mmm*) from 833 to 1073 K on heating (Figs. 2.3, 2.4, 2.5, 2.6, 2.7, and Tables 2.1 and 2.2). The 002 and 020 peaks and 602 and 620 ones of the

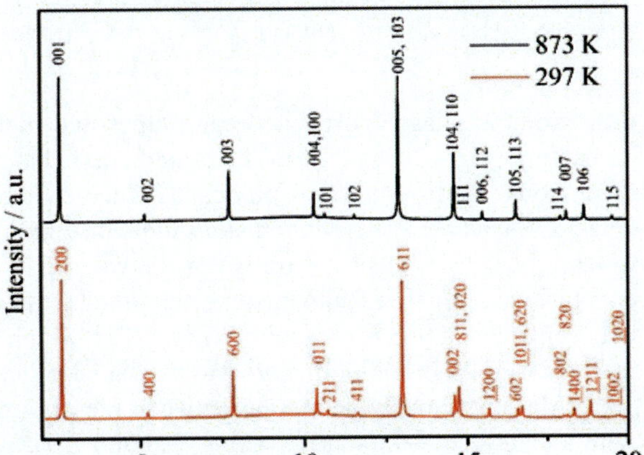

Fig. 2.3 Synchrotron X-ray powder diffraction data of $CsBi_2Ti_2NbO_{10-\delta}$ at 297 and 773 K, showing a phase transition

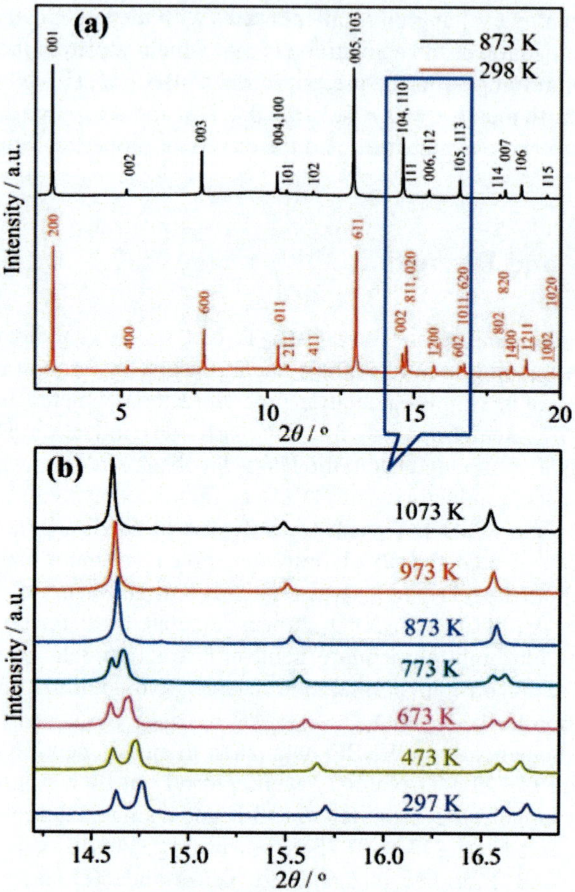

Fig. 2.4 **a** Synchrotron X-ray powder diffraction data of $CsBi_2Ti_2NbO_{10-\delta}$ measured at 297 and 773 K, showing a phase transition. **b** Inset shows the zoom-in view of synchrotron X-ray powder diffraction data between $2\theta = 14$ and 17 ° from 297 to 1073 K

orthorhombic phase approached each other and merged between 813 and 833 K on heating (Figs. 2.8 and 2.9). The *b*- and *c*-axis lengths of the orthorhombic phase continuously approached each other and coincided between 813 and 833 K on heating (Figs. 2.10 and 2.11). The conclusions indicate that there occurred an orthorhombic-to-tetragonal (o-to-t) phase transition between 813 and 833 K on heating. This conclusion is consistent with the reported literature [43]. The transition was observed between 793 and 813 K on cooling (Fig. 2.11), which exhibited a hysteresis of about 20 K (Fig. 2.12). In addition, there was a reduced lattice volume of $CsBi_2Ti_2NbO_{10-\delta}$, which discontinuously decreased between the orthorhombic and tetragonal phases with a change of −0.042(5)% (Figs. 2.8 and 2.9). Such hysteresis during heating and cooling, combined with the discontinuous decrease of lattice volume, indicated that the phase transition is a first-order phase transition. Temperature dependence

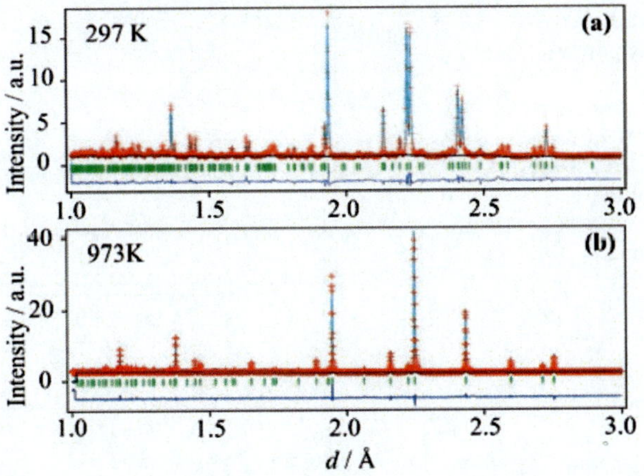

Fig. 2.5 Rietveld patterns of $CsBi_2Ti_2NbO_{10-\delta}$ at 297 K **a** and 973 K **b** of neutron-diffraction data taken by the backscattering bank ($d = 1.0–3.0$ Å) of the SuperHRPD diffractometer. The observed plots are shown by red marks, the calculated intensities are shown by light blue, and the difference plots are shown by blue solid lines. Green tick marks stand for calculated Bragg peak positions

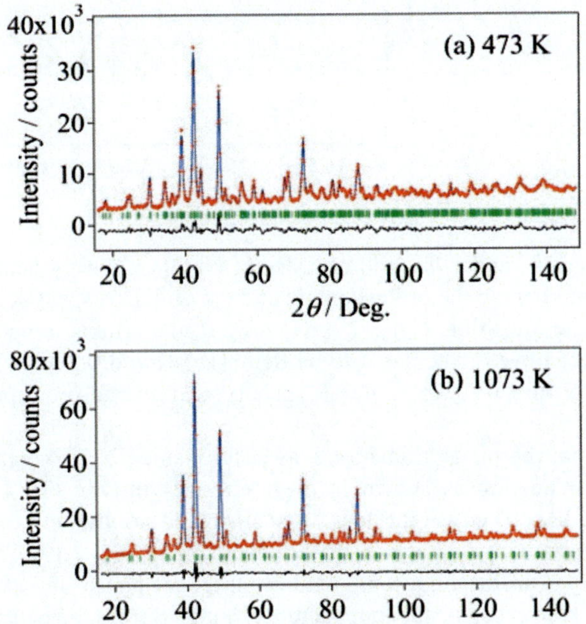

Fig. 2.6 Rietveld patterns of $CsBi_2Ti_2NbO_{10-\delta}$ at 497 K **a** and 1073 K **b** of neutron-diffraction data measured in situ at ANSTO (Echidna). The observed plots are shown by red marks, the calculated intensities are shown by light blue, and the difference plots are shown by blue solid lines. Green tick marks stand for the calculated Bragg peak positions

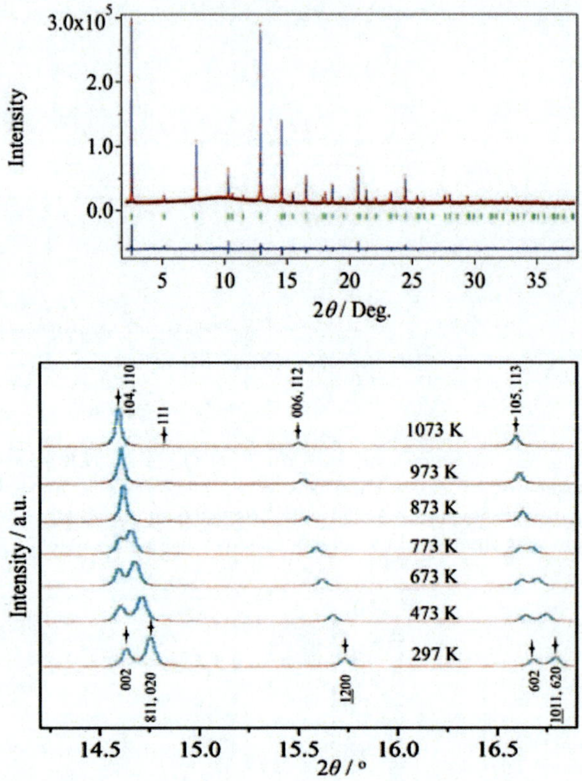

Fig. 2.7 Rietveld patterns of synchrotron X-ray powder diffraction data of $CsBi_2Ti_2NbO_{10-\delta}$ measured between 297 and 1073 K on heating. The *hkl* denotes the reflection index of the orthorhombic structure at 297 K and tetragonal structure at 1073 K

of axial ratios, c_o/b_o and a_t/b_t (=1) of $CsBi_2Ti_2NbO_{10-\delta}$ during heating where the subscripts o and t denote the orthorhombic and tetragonal, respectively. Assuming a continuous phase transition, c_o/b_o–1 ∝ $(1–T/T_C)^\eta$, the critical temperature T_C and critical exponent η were calculated to be 819.8(15) K and 0.298(10), respectively. The η was lower than 0.5, where $\eta = 0.5$ is a typical second-order phase transition (Fig. 2.10) [60].

We found that the phase transition was accompanied by a weight loss of 0.32 wt% weight on heating as shown in the thermogravimetric (TG) results (black line in Fig. 2.13). The weight loss was attributed to the increase in oxygen vacancy concentration as shown in the blue squares in Fig. 2.13. The presence of oxygen vacancy δ in the high-temperature tetragonal phase of $CsBi_2Ti_2NbO_{10-\delta}$ was confirmed by oxygen occupancy factors calculated from the neutron-diffraction data and also by the oxygen contents estimated using the TG analysis. The oxygen vacancies in the $CsBi_2Ti_2NbO_{10-\delta}$ are responsible for the good oxide-ion conductivitiy of $CsBi_2Ti_2NbO_{10-\delta}$, which will be described later. The structure of $CsBi_2Ti_2NbO_{10-\delta}$ consists of inner perovskite layer, outer perovskite layer and insulating rock-salt layer (Figs. 2.14 and 2.15).

Table 2.1 Crystallographic parameters of $CsBi_2Ti_2NbO_{9.97(1)}$ which were refined using neutron-diffraction data taken at 297 K with SuperHRPD diffraction

Site label s	Atom X oxidation number		$g(X; s)^a$	x	y	z	U_{eq} (Å^2)^b	BVS^c
Cs	Cs + 1		1.000^d	1/4	0.49941(4)	0.0012(4)	0.0126(4)	0.98
Bi	Bi + 3		0.992(2)	0.92575(3)	0.50806(18)	−0.0342(1)	0.0250(2)	2.97
Ti/Nb1	Ti	+ 4.18	0.822(2)^e	0	0	−0.01774(13)	0.0059(1)	4.44
	Nb		0.178(2)^e					
Ti/Nb2	Ti	+ 4.41	0.589(1)^e	0.85896(2)	0.00575(18)	0.04243(18)	0.0034(2)	4.18
	Nb		0.411(1)^e					
O1	O − 2		0.994(2)	0.0055(9)	0.75929(4)	0.28819(4)	0.0410(4)	2.04
O2	O − 2		1.0000^d	0.937(5)	−0.05406(2)	0.05269(3)	0.0299(3)	1.99
O3	O − 2		1.0000^d	0.87042(7)	0.77108(4)	0.30977(3)	0.0296(4)	2.08
O4	O − 2		0.998(2)	0.37499(5)	0.73288(3)	0.26039(3)	0.0085(3)	2.08
O5	O − 2		0.995(2)	0.80157(4)	0.02197(4)	0.01392(2)	0.0089(3)	1.75

Crystal system: orthorhombic; Space group: *Ima*2; Lattice parameters: $a = 30.76478(10)$ Å, $b = 5.450391(14)$ Å, and $c = 5.498837(14)$ Å, which is consistent with the literature [2]. Reliability factors: $R_{wp} = 6.06\%$, $R_p = 4.48\%$, $R_B = 6.33\%$, and $R_F = 5.99\%$

[a] $g(X; s)$: Occupancy factor of X atom at the s site

[b] U_{eq}: Equivalent isotropic atomic displacement parameter

[c] Bond-valence sums (BVSs) were calculated using bond-valence parameters from Ref. [3, 4]

[d] The occupancy factors of Cs, Bi, and O atoms were refined in preliminary analyses. The refined occupancy factors of Cs, O3, and O4 sites were fixed at 1 because their occupancies were higher than 1

[e] The following linear constraints were used in the refinement: g(Ti; Ti/Nb1) + g(Nb; Ti/Nb1) = 1, g(Ti; Ti/Nb2) + g(Nb; Ti/Nb2) = 1, [g(Ti; Ti/Nb1) + g(Ti; Ti/Nb2)] : [g(Nb; Ti/Nb1) + g(Nb; Ti/Nb2)] = 2:1

Table 2.2 Crystallographic parameters of $CsBi_2Ti_2NbO_{9.80(2)}$ which were refined using neutron-diffraction data taken with SuperHRPD diffraction at 973 K

Site label *s*	Atom *X* oxidation number		*g*(*X*; *s*)	*x*	*y*	*z*	U_{eq} (Å^2)	BVS[c]
Cs	Cs + 1		1.000[a]	1/2	1/2	1/2	0.0472(8)	0.84
Bi	Bi + 3		1.000[a]	1/2	1/2	0.1485(1)	0.0789(5)	2.49
Ti/Nb1	Ti	+ 4.20	0.804(2)[b]	0	0	0	0.0075[d]	4.34
	Nb		0.196(2)[b]					
Ti/Nb2	Ti	+ 4.40	0.5978(11)[b]	0	0	0.2804(6)	0.0075[d]	4.31
	Nb		0.4022(11)[b]					
O1	O − 2		0.996(3)	0	1/2	0	0.0879(9)	1.84
O2	O − 2		0.976(3)	0	0	0.1257(1)	0.0734(8)	1.72
O3	O − 2		0.991(1)	0	1/2	0.2541(9)	0.0441(5)	1.99
O4	O − 2		0.944(3)	0	0	0.3926(1)	0.0382(5)	1.77

Site *s*	U_{11} (Å^2)[c]	U_{22} (Å^2)[c]	U_{33} (Å^2)[c]	U_{12} (Å^2)	U_{13} (Å^2)[c]	U_{23} (Å^2)[c]
Cs	0.0248(9)	0.0248(9)	0.0918(3)	0	0	0
Bi	0.0650(6)	0.0650(6)	0.1068(3)	0	0	0
Ti/Nb1	0.0075[d]	0.0075[d]	0.0075[d]	0	0	0
Ti/Nb2	0.0075[d]	0.0075[d]	0.0075[d]	0	0	0
O1	0.0901(19)	0.0170(14)	0.157(3)	0	0	0
O2	0.1010(11)	0.1010(11)	0.0181(17)	0	0	0
O3	0.0100(4)	0.0100(4)	0.1123(18)	0	0	0
O4	0.0559(8)	0.0559(8)	0.0026(14)	0	0	0

Crystal system: tetragonal; Space group: *P4/mmm*; Lattice parameters: $a = b = 3.894032(7)$ Å and $c = 15.58224(6)$ Å, which are consistent with the literature [2]. However, M
Reliability factors: $R_{wp} = 2.39\%$, $R_p = 1.78\%$, $R_B = 4.05\%$, and $R_F = 8.01\%$

[a] The occupancy factors of Cs, Bi, and O atoms were refined in preliminary analyses. The occupancy factors were refined to be *g*(Cs; Cs) = 1.009(3), *g*(Bi; Bi) = 1.002(2), which indicates that *g*(Cs; Cs) = *g*(Bi; Bi) = 1.000 within 3 times of the estimated deviation

[b] The following linear constraints were used in the refinement:

$$g(\text{Ti; Ti/Nb1}) + g(\text{Nb; Ti/Nb1}) = 1, g(\text{Ti; Ti/Nb2}) + g(\text{Nb; Ti/Nb2}) = 1, [g(\text{Ti; Ti/Nb1}) + g(\text{Ti; Ti/Nb2})] : [g(\text{Nb; Ti/Nb1}) + g(\text{Nb; Ti/Nb2})] = 2 : 1$$

[c] U_{ij}: Anisotropic atomic displacement parameter

[d] Atomic displacement parameters of Ti and Nb atoms were refined in preliminary analyses using the following linear constraints: U_{eq}(Ti/Nb1) = U_{eq}(Ti/Nb2) and fixed to the value of isotropic atomic displacement parameter (U_{iso}) in the final refinement

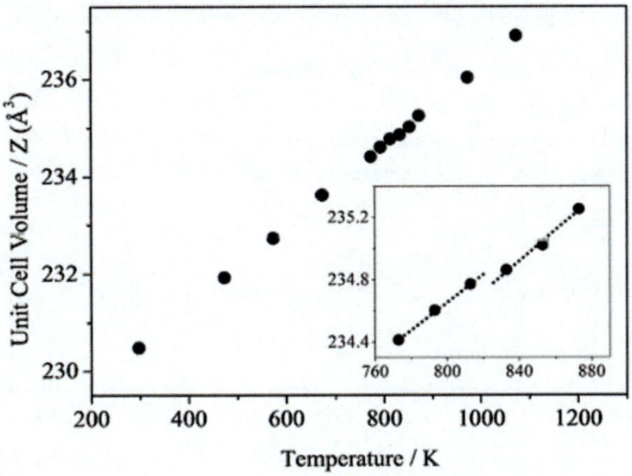

Fig. 2.8 Reduced lattice volume of $CsBi_2Ti_2NbO_{10-\delta}$, refined by synchrotron X-ray powder diffraction (SXRPD) data

The electrical conductivities measurements of $CsBi_2Ti_2NbO_{10-\delta}$ were carried out in the temperature ranges of 573 to 1173 K using AC and DC methods [68]. The bulk and grain boundary responses were observed and evidenced at high frequencies and equivalent circuit analyses were carried out to extract the bulk conductivity σ_b and grain boundary conductivity σ_{gb} at each temperature range using the ZView (Fig. 2.16). Figures 2.17, 2.18 and 2.19 show the AC impedance spectra of $CsBi_2Ti_2NbO_{10-\delta}$. Unfortunately, the electrode component was not observed clearly at relatively low frequencies. However, two distinct arcs for the bulk responses and for the grain boundary responses were clearly obtained in the complex plane plots. The bulk and grain boundary responses were observed in the higher frequency regions with respective capacitance values of $C_b \approx 0.7 \times 10^{-12}$ F cm^{-1} and $C_{gb} \approx 7.0 \times 10^{-9}$ F cm^{-1}. The bulk conductivities σ_b were independent of different oxygen partial pressure of O_2, air, and N_2; gases (Figs. 2.20 and 2.21), which indicated the ion conductivity in the $CsBi_2Ti_2NbO_{10-\delta}$. Figure 2.22 shows the temperature dependence of the bulk conductivities σ_b and grain boundary conductivities σ_{gb} of $CsBi_2Ti_2NbO_{10-\delta}$ on heating. We found that the σ_b increased between 673 and 873 K on heating abruptly, which can be attributed to the abrupt increase of the conduction carrier of oxygen vacancy concentration and also the phase transition from orthorhombic to tetragonal phases. The total DC electrical conductivities of $CsBi_2Ti_2NbO_{10-\delta}$ also exhibited a similar abrupt increase around the transition point (Figs. 2.23 and 2.24). The abrupt increase in σ_{tot} at 873 K is attributed not to the increase of n-type conductivity, but to the increase of oxide-ion conductivity, because the transport number in O_2/air was 1.00 at 873 K. The increased oxide-ion conductivities are ascribed to the increase in oxygen vacancies concentration where the formation of the oxygen vacancies could be explained by the following defect reaction equations.

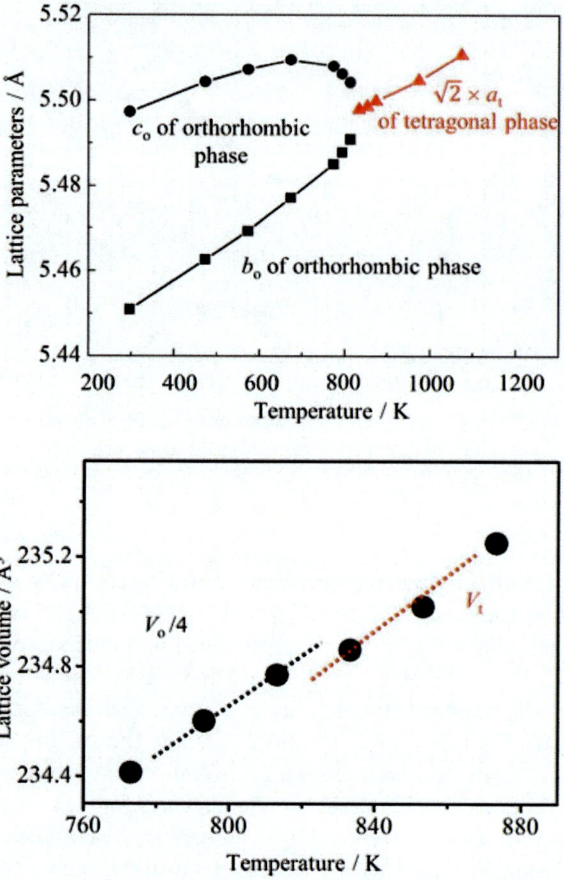

Fig. 2.9 Lattice parameters **a** and reduced lattice volume **b** of $CsBi_2Ti_2NbO_{10-\delta}$, which were refined using synchrotron X-ray powder diffraction data measured in situ at high temperatures in air on heating. The subscripts o and t denote orthorhombic and tetragonal, respectively

$$O_O^x + Ti^{4+} \rightarrow \frac{1}{2}O_2(g) + V_{\ddot{O}} + Ti^{3+}$$

$$\text{and/or } O_O^x + Nb^{5+} \rightarrow \frac{1}{2}O_2(g) + V_{\ddot{O}} + Nb^{4+}$$

$$\text{and/or } O_O^x + Nb^{4+} \rightarrow \frac{1}{2}O_2(g) + V_{\ddot{O}} + Nb^{3+}$$

Oxygen concentration-cell measurements were carried out to determine the t_{ion} of $CsBi_2Ti_2NbO_{10-\delta}$. As the relative density of the pellet used here was 91%, we used a thick pellet (4.5 mm) in order to avoid gas penetration between these two sides. The t_{ion} was 1.00−0.98 between 873 and 1173 K in air/O_2, 0.97−0.95 between 873

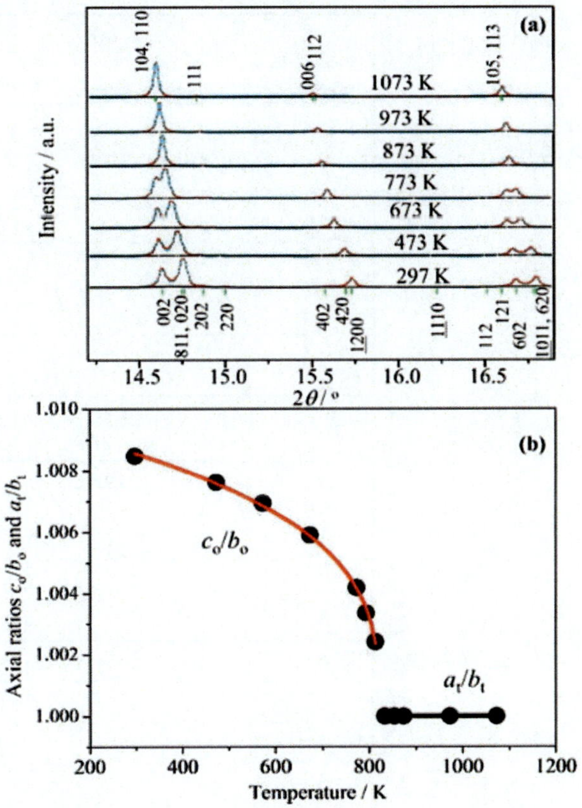

Fig. 2.10 **a** Rietveld patterns of synchrotron X-ray powder diffraction data of $CsBi_2Ti_2NbO_{10-\delta}$ measured between 297 and 1073 K on heating. The *hkl* denotes the reflection index of the orthorhombic structure at 297 K and tetragonal structure at 1073 K. **b** Temperature dependence of axial ratios, c_o/b_o and a_t/b_t (= 1) of $CsBi_2Ti_2NbO_{10-\delta}$ during heating where the subscripts o and t denote the orthorhombic and tetragonal, respectively

and 1173 K in air/N_2, and 0.87 at 873 K in air/5% H_2 in N_2 (Figs. 2.25, 2.26, and 2.27). Total DC electrical conductivity σ_{tot} at 973 K was almost independent of the oxygen partial pressure $P(O_2)$ between $P(O_2) = 2.0 \times 10^{-22}$ and 1 atm (Figs. 2.28 and 2.29). At $P(O_2) < 2.0 \times 10^{-22}$ region, an n-type electronic contribution to the total conductivity was observed, which is consistent with the lower t_{ion} in air/5% H_2 in N_2. The electronic conduction is attributable to the reduction from Ti^{4+} and/or Nb^{5+} cations leading to the formation of oxygen vacancies and electronic defects as follows.

$$O_O^X + \delta Ti^4 \rightarrow \frac{1}{2}O_2(g) + V_{\ddot{O}} + \delta Ti^{3+} + (2-\delta)e'$$

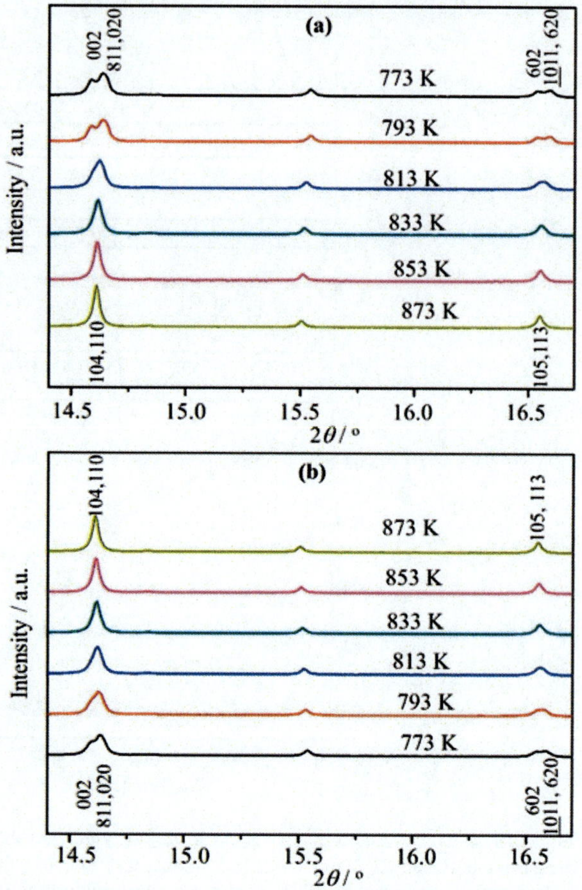

Fig. 2.11 Synchrotron X-ray diffraction patterns from 773 to 873 K on heating **a** and cooling **b** processes, showing a hysteresis of about 20 K

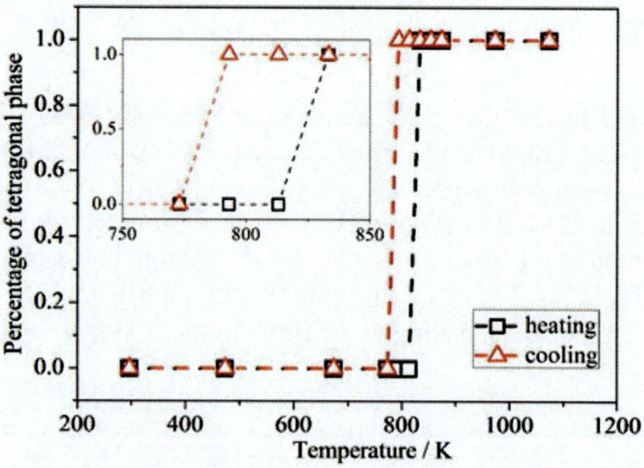

Fig. 2.12 Percentage of tetragonal phase from SXRPD results, showing a hysteresis

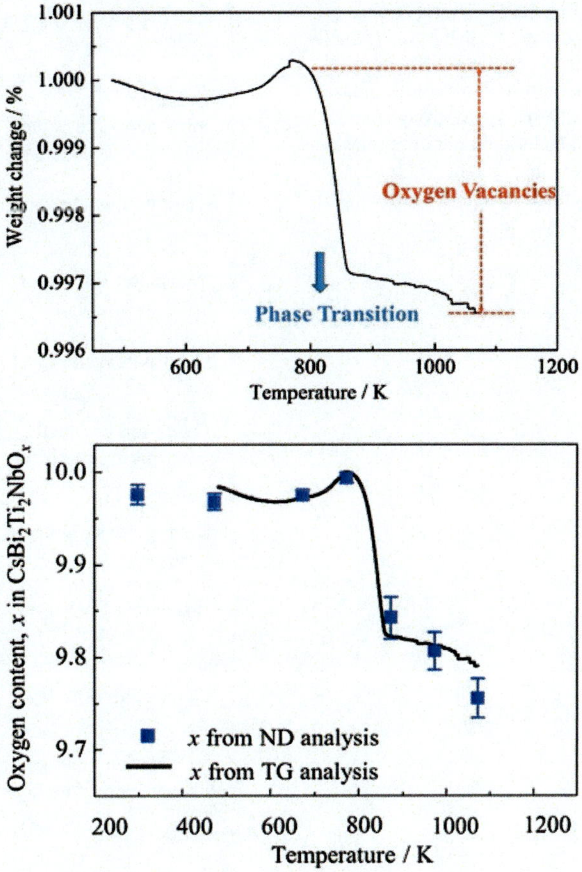

Fig. 2.13 Oxygen contents of $CsBi_2Ti_2NbO_{10-\delta}$ on heating, which were obtained by thermogravimetric analysis (black line) and calculated from occupancy factors refined using neutron-diffraction data (blue marks)

$$\text{and/or } O_O^x + \delta Nb^{5+} \rightarrow \frac{1}{2}O_2(g) + V_{\ddot{O}} + \delta Nb^{4+} + (2-\delta)e'$$

$$\text{and/or } O_O^x + \delta\, Nb^{4+} \rightarrow \frac{1}{2}O_2(g) + V_{\ddot{O}} + \delta Nb^{3+} + (2-\delta)e'$$

We cannot observe significant proton conductivities between 873 and 1173 K ranges, because the total DC conductivities measured in dry air have been agreed well with those measured in wet air (Fig. 2.30). The H_2O partial pressure of dry air is lower than 1.8×10^{-4} atm and the H_2O partial pressure of wet air is 2.3×10^{-2} atm. Therefore, we can get the conclusion that the oxide ions are the dominant conduction carrier in the $CsBi_2Ti_2NbO_{10-\delta}$ and that the $CsBi_2Ti_2NbO_{10-\delta}$ is a pure oxide-ion conductor. There was no change observed when we compared the X-ray powder diffraction patterns before and after impedance spectroscopy and also before and after the oxygen concentration-cell measurements (Fig. 2.31), which indicated the good phase stability of $CsBi_2Ti_2NbO_{10-\delta}$ at different temperature ranges and

Fig. 2.14 Crystal structures of $CsBi_2Ti_2NbO_{10-\delta}$ refined using neutron-diffraction data: low-temperature phase (297 K, space group: *Ima*2). The low-temperature phase exhibits octahedral tilting and polar displacements [12]

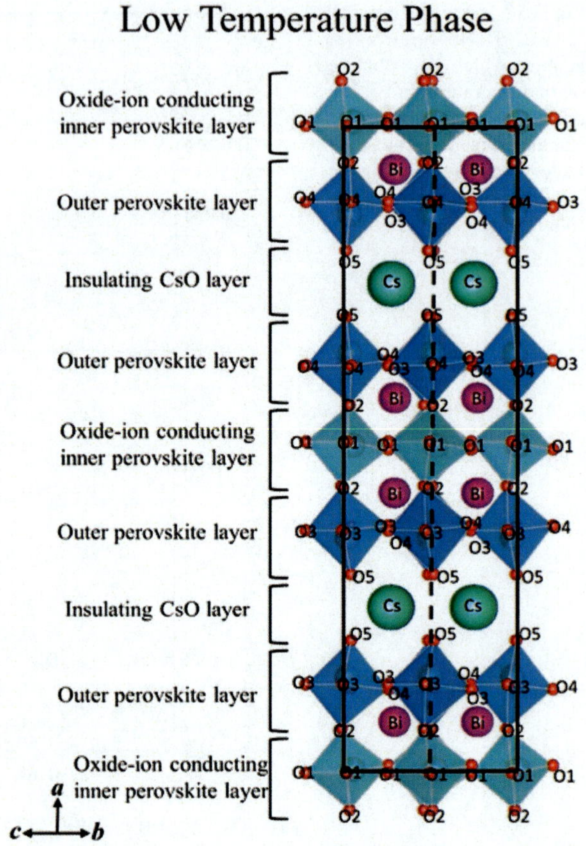

different $P(O_2)$ ranges. The σ_b of $CsBi_2Ti_2NbO_{10-\delta}$ was higher than that of YSZ at high-temperature ranges and was comparable to those of the best oxide-ion conductors (Fig. 2.32) [69–72]. This indicates the high potential of Dion–Jacobson phase $CsBi_2Ti_2NbO_{10-\delta}$ as a basic composition for the oxide-ion conductors.

On the other hand, we have discussed the structure origin of the novel oxide-ion conductivity using the refined structure of $CsBi_2Ti_2NbO_{9.80(2)}$ at 973 K (Fig. 2.15). The crystal structure was obtained using the Rietveld refinement by the neutron-diffraction data, which was measured in situ at 973 K using SuperHRPD [59]. The high-temperature structure of tetragonal $CsBi_2Ti_2NbO_{10-\delta}$ consists of oxide-ion conducting inner perovskite $Bi(Ti_{0.804}Nb_{0.196})O_{3-3\delta/10}$ layer, outer perovskite $Bi(Ti_{0.5975}Nb_{0.4025})O_{3-3\delta/10}$ layer, and insulating $CsO_{1-\delta/10}$ rock-salt layer (Figs. 2.14 and 2.15). The refined equivalent isotropic atomic displacement parameters of equatorial oxygen O1, $U_{eq}(O1)$, in the oxide-ion conducting inner perovskite layer and apical oxygen O2, $U_{eq}(O2)$ in the oxide-ion conducting inner perovskite layer were much higher than those of equatorial O3 in the outer perovskite layers and apical

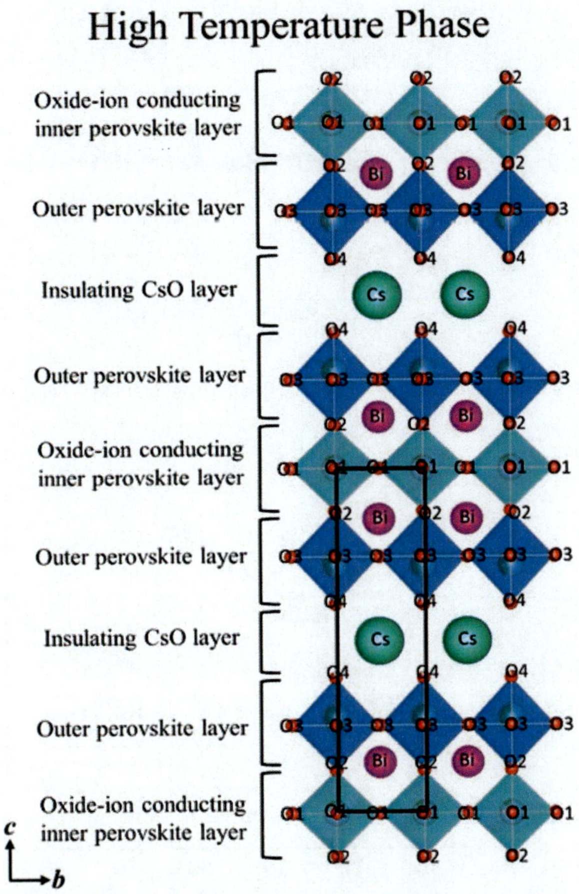

Fig. 2.15 Crystal structures of $CsBi_2Ti_2NbO_{10-\delta}$ refined using neutron-diffraction data: high-temperature phase (973 K, space group: *P4/mmm*)

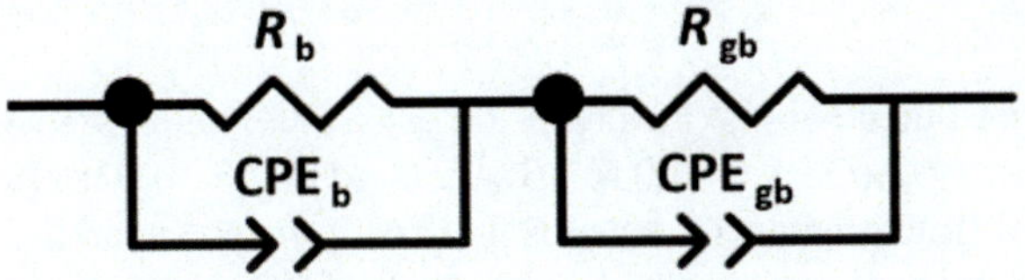

Fig. 2.16 Equivalent circuit used to fit the impedance spectroscopy data. *R* denotes a resistor and CPE stands for a constant phase element, the subscripts b and gb stand for bulk and grain boundary, respectively

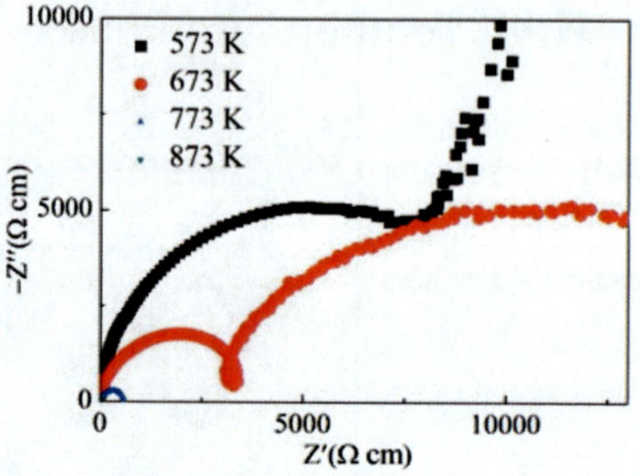

Fig. 2.17 Complex impedance plots of $CsBi_2Ti_2NbO_{10-\delta}$ recorded in flowing dry air from 573 to 873 K on heating

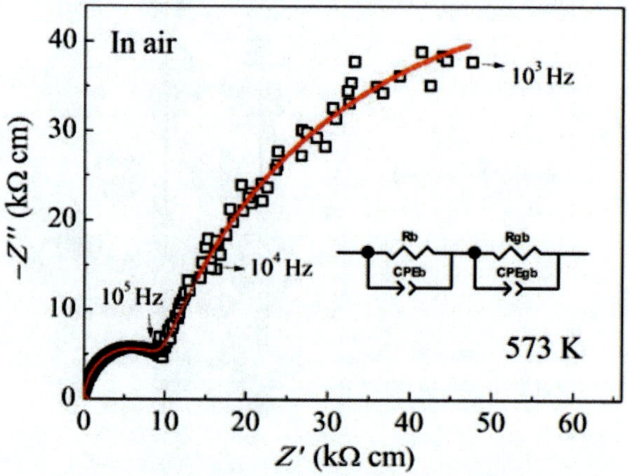

Fig. 2.18 Complex impedance plots of $CsBi_2Ti_2NbO_{10-\delta}$ recorded in flowing dry air at 573 K on heating. Red line indicates the equivalent circuit fitting

O4 atoms in the outer perovskite layers ($U_{eq}(O1) = 0.0879(9)Å^2$, $U_{eq}(O2) = 0.0734(8)Å^2 >> U_{eq}(O3) = 0.0441(5)Å^2$, $U_{eq}(O4) = 0.0382(5)Å^2$). We used three different diffractometers (a. a neutron diffractometer SuperHRPD at the Materials and Life Science Experimental Facility of J-PARC, Japan, b. a neutron diffractometer Echidna at the Australian Nuclear Science and Technology Organization, Australia (ANSTO). c. a synchrotron X-ray powder diffractometer at the BL02B2 beam line of SPring-8, Japan). All the three data showed similar results that the O1 and O2 atoms show externally large thermal parameters than others. In our work, we used the data taken from J-PARC because the SuperHRPD diffractometer has the world's best resolution of $\Delta d/d = 0.035\%$ in the lattice space d scale. Therefore, the O1 and O2 atoms in the inner oxide-ion conducting perovskite layer exhibited larger thermal motions than the O3 and O4 atoms in the outer perovskite layer. In

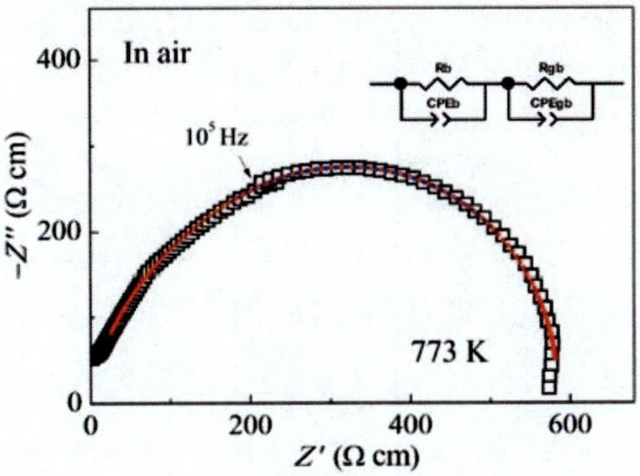

Fig. 2.19 Complex impedance plots of $CsBi_2Ti_2NbO_{10-\delta}$ recorded in flowing dry air at 773 K on heating. Red line indicates the equivalent circuit fitting

Fig. 2.20 Complex impedance plots of $CsBi_2Ti_2NbO_{10-\delta}$ measured in flowing dry N_2, air, and O_2 gases at 673 K

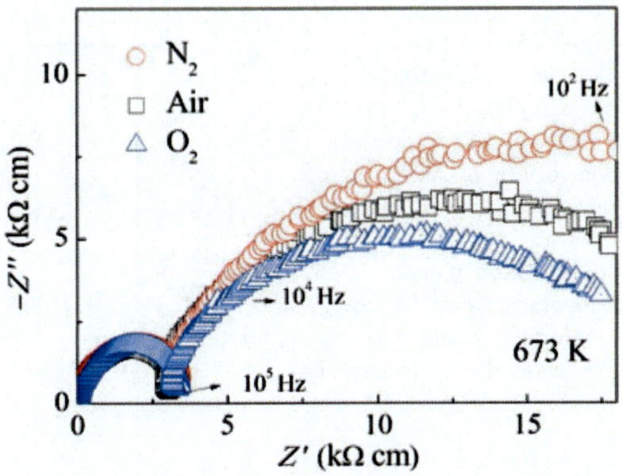

Fig. 2.21 Complex impedance plots of $CsBi_2Ti_2NbO_{10-\delta}$ measured in flowing dry N_2, air, and O_2 gases at 973 K

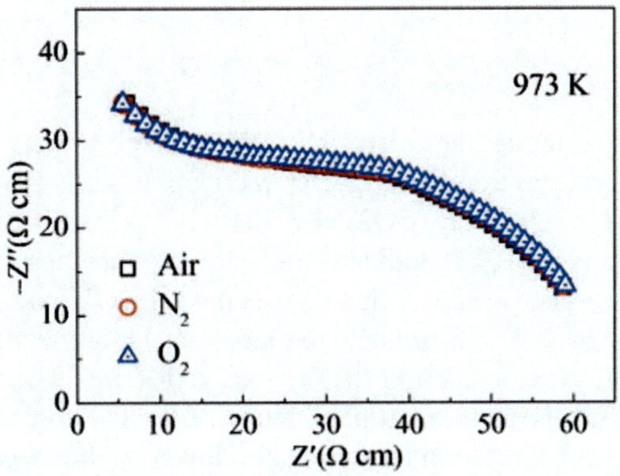

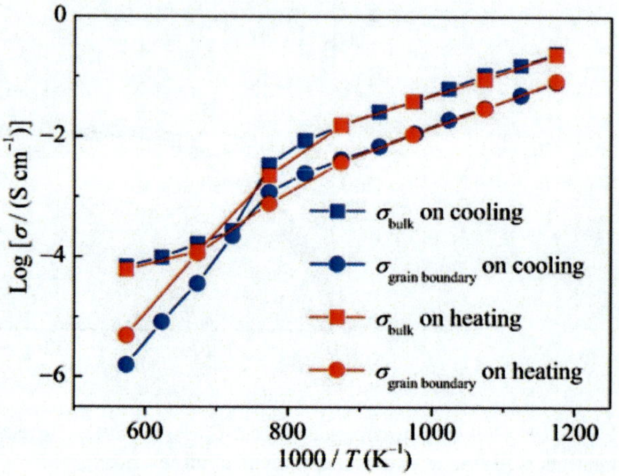

Fig. 2.22 Bulk electrical conductivity σ_b and grain boundary conductivity σ_{gb} of $CsBi_2Ti_2NbO_{10-\delta}$ in air on heating

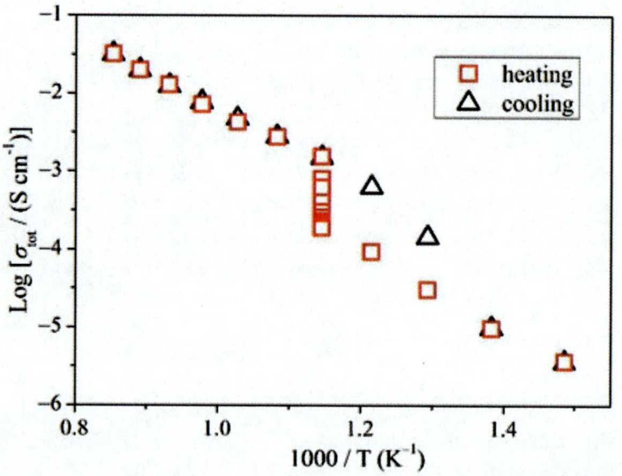

Fig. 2.23 Total DC electrical conductivities σ_{tot} of $CsBi_2Ti_2NbO_{10-\delta}$ from 673 to 1173 K on heating and cooling processes in air, measured every 50 K and allowed for 1 h of equilibration at each temperature step. The σ_{tot} on heating and cooling processes show a hysteresis

particular, the anisotropic atomic displacement parameters (U_{ij}) of O1 and O2 atoms were extremely large: $U_{11}(O1) = 0.0901(19)$ Å^2, $U_{33}(O1) = 0.1567(27)$ Å^2, and $U_{11}(O2) = U_{22}(O2) = 0.1010(11)$ Å^2 (see others in Table 2.2). The large $U_{33}(O1)$ and $U_{11}(O2)$ indicate high anisotropic thermal motions along the c- and a-axes, respectively, which suggests the O1−O2 oxide-ion diffusion (blue dotted arrows in Fig. 2.33). Similarly, the large $U_{11}(O1)$ means high thermal motions along a- and b-axes, indicating the O1−O1 oxide-ion diffusion (red dotted arrows in Fig. 2.33). The large thermal parameters of the O1 and O2 are attributable to the large bottlenecks in the inner perovskite layer by the large size of Cs^+ and Bi^{3+} displacement. Isosurfaces of neutron scattering length density obtained by the maximum-entropy method (MEM) [14, 15, 64, 73–77] (yellow isosurfaces in Fig. 2.34) and bond-valence-based energy landscapes (BVELs, blue isosurfaces in Fig. 2.35) also showed anisotropic thermal motions of O1 and O2 atoms. The O1−O2 and O1−O1 oxide-ion

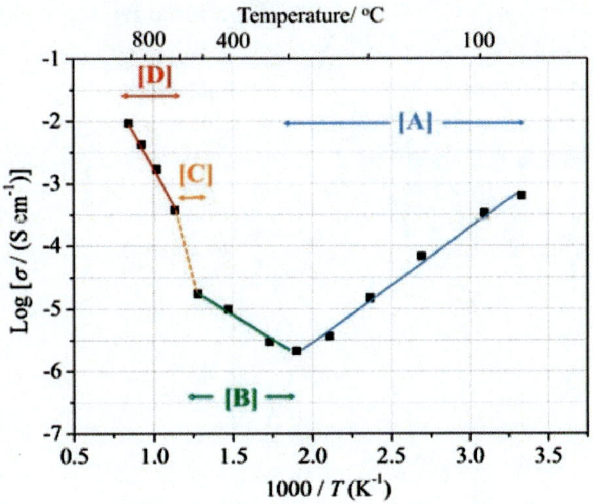

Fig. 2.24 Arrhenius plots of σ total electrical conductivity σ_{tot} of $CsBi_2Ti_2NbO_{10-\delta}$ from room temperature to 1073 K

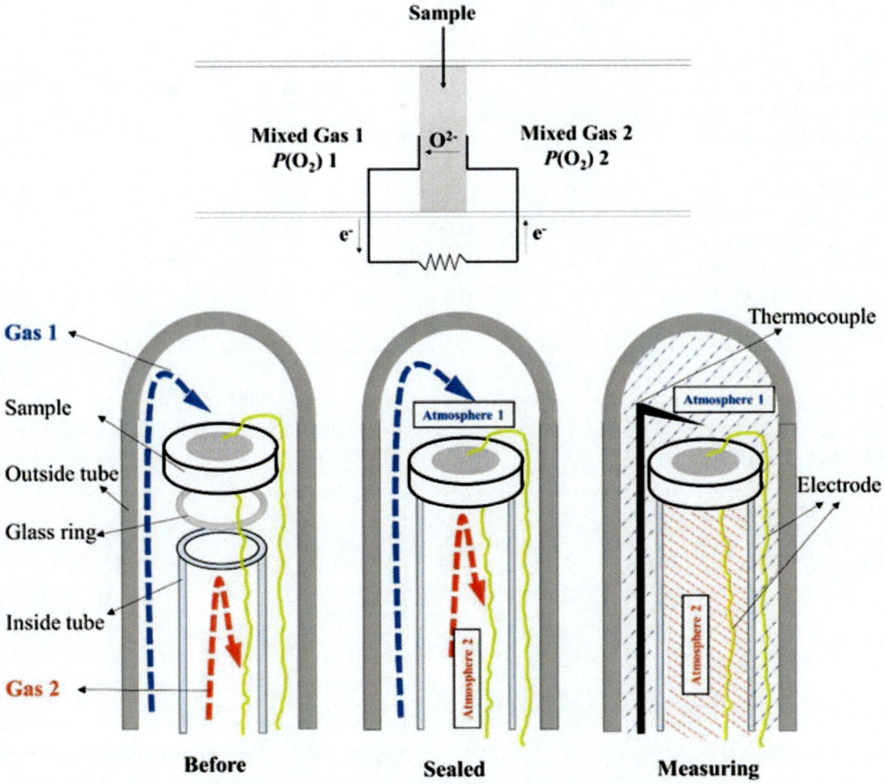

Fig. 2.25 Schematic representation of oxygen concentration-cell measurements

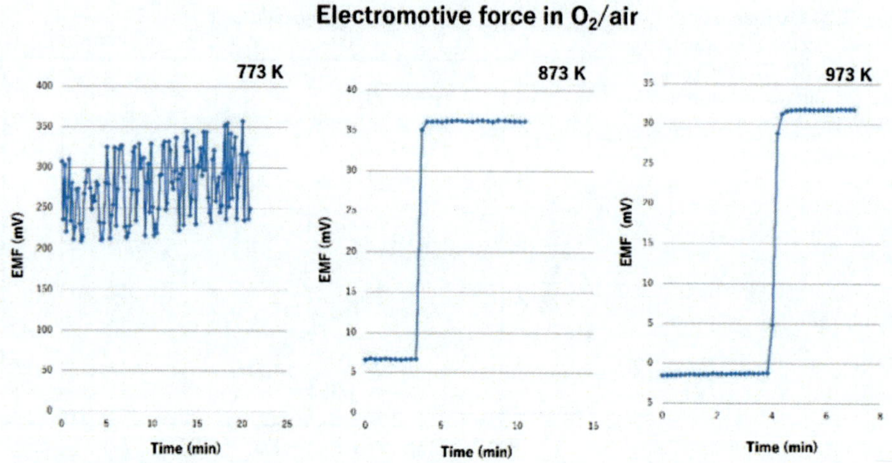

Fig. 2.26 Electromotive force in O_2/air condition at 773 K. 873 and 973 K using oxygen concentration-cell measurement

Fig. 2.27 Oxide-ion transport number t_{ion} determined by oxygen concentration cell measurements of $CsBi_2Ti_2NbO_{10-\delta}$ by flowing dry air and the other side to a flowing dry O_2 (Air/O_2), N_2 (Air/N_2), and 5% H_2 in N_2 (Air/5% H_2 in N_2) gases

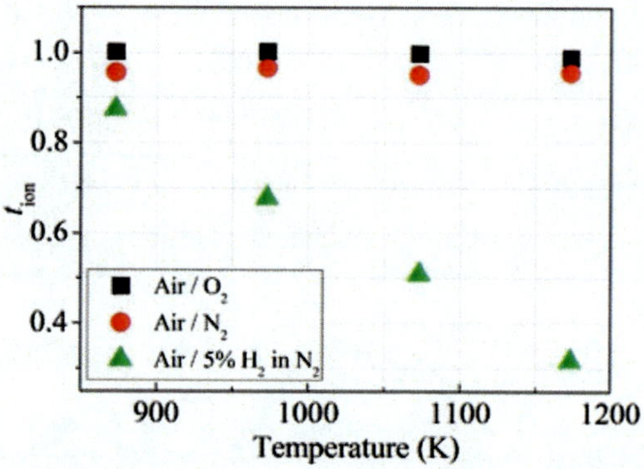

Fig. 2.28 Partial oxygen pressure $P(O_2)$ dependence of total DC electrical conductivities σ_{tot} of $CsBi_2Ti_2NbO_{10-\delta}$ at 1073 K

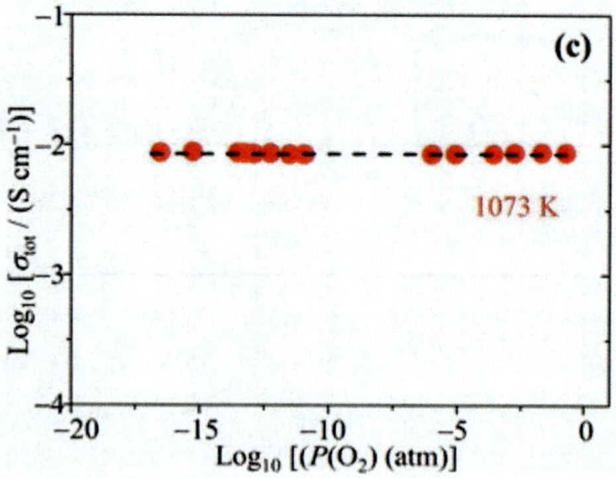

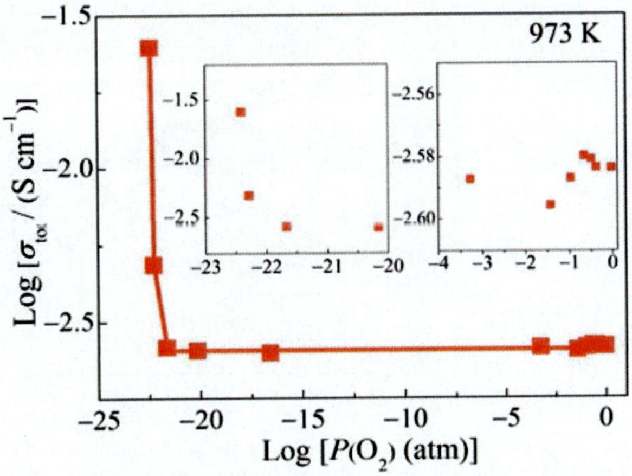

Fig. 2.29 Partial oxygen pressure $P(O_2)$ dependence of total DC electrical conductivities σ_{tot} of $CsBi_2Ti_2NbO_{10-\delta}$ at 973 K

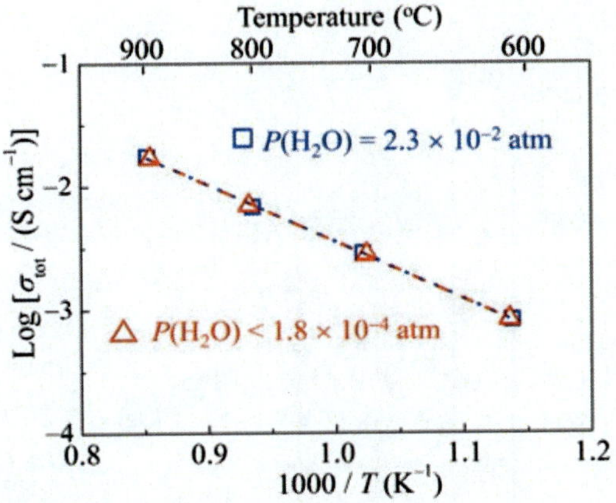

Fig. 2.30 Arrhenius plots of σ_{tot} of $CsBi_2Ti_2NbO_{10-\delta}$ in dry and wet atmospheres

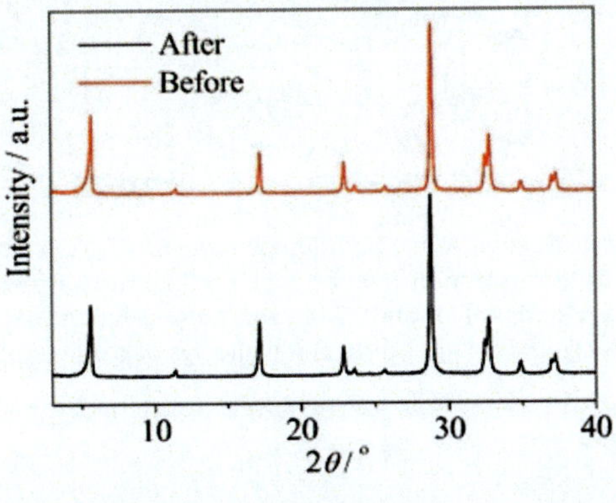

Fig. 2.31 X-ray powder diffraction patterns before and after impedance spectroscopy and oxygen concentration-cell measurements

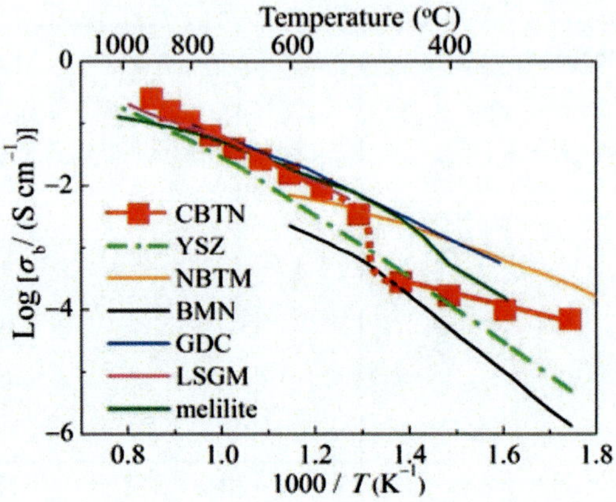

Fig. 2.32 Comparison of bulk conductivity σ_b of $CsBi_2Ti_2NbO_{10-\delta}$ (CBTN) on cooling and best oxide-ion conductors

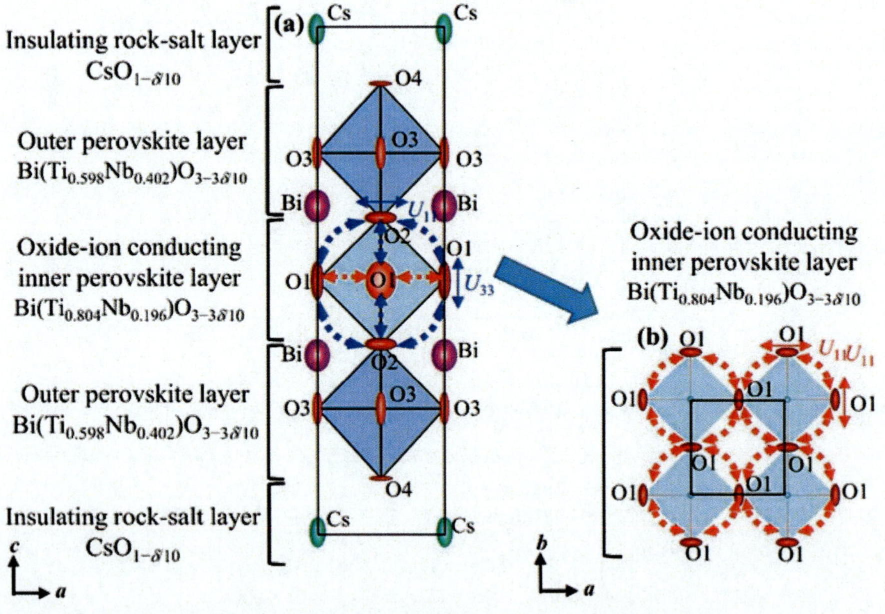

Fig. 2.33 Refined crystal structure of the high-temperature $CsBi_2Ti_2NbO_{9.80(2)}$ at 973 K, refined using Rietveld analyses by the neutron-diffraction data. The lines with arrows stand for the directions of the anisotropic thermal motions of O1 and O2 oxygen atoms. Blue and red dotted lines with arrows denote the possible O1−O2 and O1−O1 diffusion pathways of oxide ions, respectively

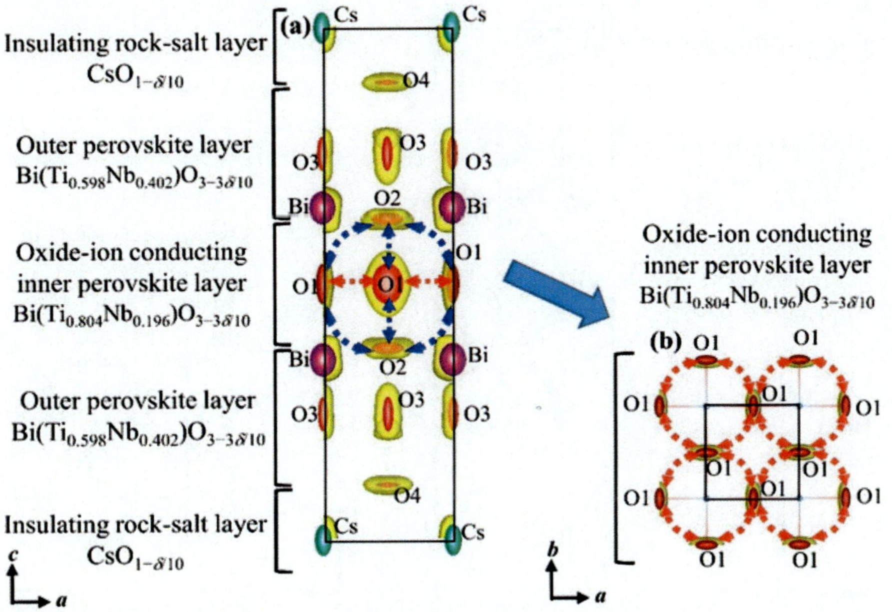

Fig. 2.34 Yellow isosurfaces of the neutron scattering length density at 1.0 fm Å^{-3} with the structure at 973 K. Blue and red dotted lines with arrows denote the possible O1−O2, O1−O1 diffusion pathways of oxide ion, respectively. Thermal ellipsoids are drawn at the 50% probability level

diffusion paths were clearly observed in the BVELs (blue and red dotted arrows in Fig. 2.35. In addition, the O2−O3 diffusion path was observed in the BVELs (black dotted arrows in Fig. 2.36). These results indicate two-dimensional (2D) oxide-ion diffusion along the edges of $Bi(Ti_{0.804}Nb_{0.196})O_{3-3\delta/10}$ octahedron in the oxide-ion conducting inner perovskite layer. We attribute the high oxide-ion conductivity of $CsBi_2Ti_2NbO_{10-\delta}$ not only to the existence of oxygen vacancies but also to the 2D oxide-ion diffusion and the extremely high U_{eq}(O1), U_{eq}(O2), U_{11}(O1), U_{33}(O1) and U_{11}(O2) [= U_{22}(O2)]. In fact, these thermal parameters of $CsBi_2Ti_2NbO_{10-\delta}$ are much higher than those of perovskite-type and layered perovskite-type conductors such as $La_{0.9}Sr_{0.1}Ga_{0.79}Mg_{0.21}O_{2.82}$, [78] $La_{0.8}Sr_{0.2}Ga_{0.8}Mg_{0.15}Co_{0.05}O_{2.8}$, [79] $BaBi_4Ti_4O_{15}$, [23] $Bi_4Ti_3O_{12}$, [80] and $Bi_2Sr_2TiNb_2O_{12}$ [81] (Tables 2.3 and 2.4).

Finally, we propose a new concept, large bottlenecks for oxide-ion migration by the large size of Cs^+ and Bi^{3+} displacement. The bottlenecks in $CsBi_2Ti_2NbO_{10-\delta}$ are the Bi−Bi−Ti triangles (yellow and pink triangles in Figs. 2.37 and 2.38). The large bottleneck sizes of $CsBi_2Ti_2NbO_{10-\delta}$ can be explained using the new concept as described below. The Cs^+ expands the Bi–Bi distance along the *b*-axis (orange arrows in Fig. 2.37) due to the cation ordering of larger $Cs^+$and smaller Bi^{3+}. The Bi^{3+} cation is displaced along the *c*-axis apart from the Ti/Nb1−O1 layer, which can be explained by the electrostatic forces (Fig. 2.39). The Bi^{3+} cations are displaced apart from the Ti/Nb1−O1 layer. The displacement of a Bi^{3+} cation at $z = 0.14851$ along the *c*-axis from the center of Ti/Nb1 site ($z = 0$) and Ti/Nb2 sites ($z = 0.2804$)

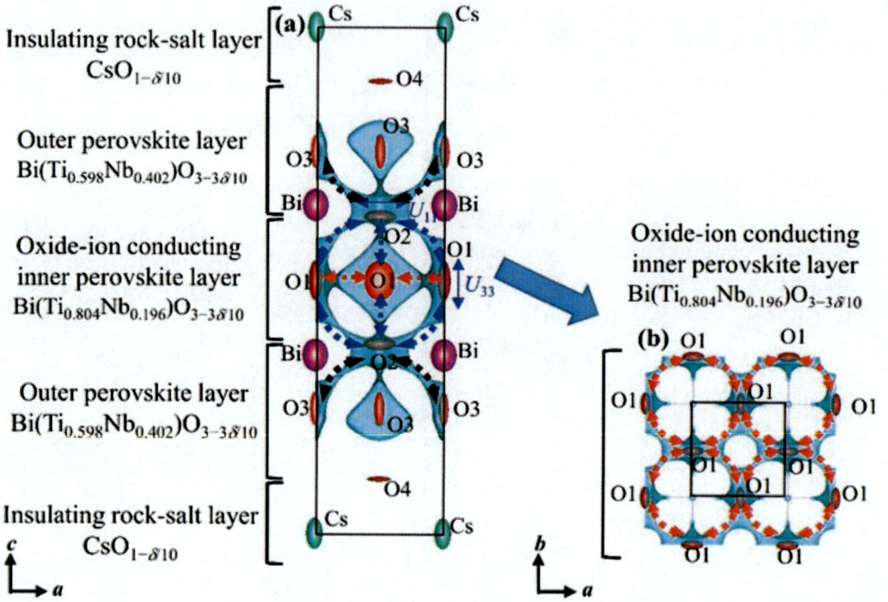

Fig. 2.35 Blue isosurfaces of the bond-valence-based energy for an oxide ion at 0.6 eV for the structure at 973 K. Blue, red, and black dotted lines with arrows denote the possible O1−O2, O1−O1, and O2−O3 diffusion pathways of oxide ion, respectively. Thermal ellipsoids are drawn at the 50% probability level

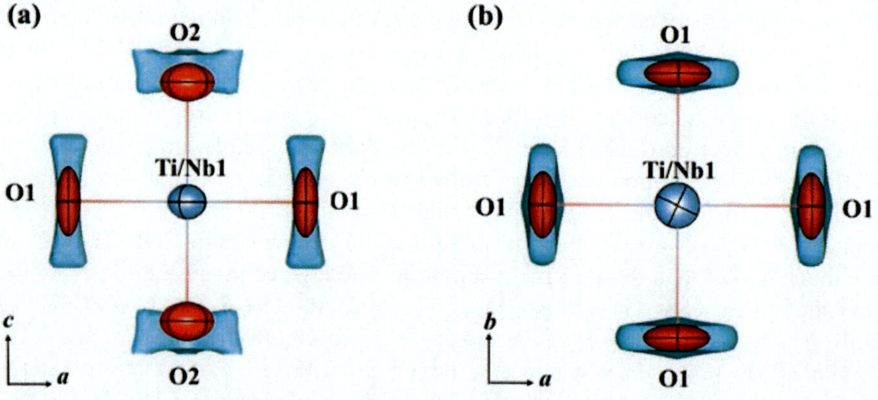

Fig. 2.36 Bond-valence-based energy landscapes (BVELs) for an oxide ion along **a** the *b*- and **b** *c*-axes with the blue isosurface at 0.35 eV and red thermal ellipsoids, for the refined crystal structure of $CsBi_2Ti_2NbO_{9.80(2)}$ at 973 K. Thermal ellipsoids are drawn at the 50% probability level. The BVELs are consistent with the shapes of thermal ellipsoids and anisotropic thermal motion of O1 and O2 atoms

Table 2.3 Comparisons of atomic displacement parameters of $CsBi_2Ti_2NbO_{10-\delta}$ with those of perovskite-type oxide-ion conductors

Composition	$CsBi_2Ti_2NbO_{10-\delta}$			$La_{0.9}Sr_{0.1}Ga_{0.79}Mg_{0.21}O_{2.82}$			$La_{0.8}Sr_{0.2}Ga_{0.8}Mg_{0.15}Co_{0.05}O_{2.8}$		
Temperature	973 K			1023 K			1069 K		
U_{ij} (Å^2)	O1	U_{11}	0.0901(19)	O	U_{11}	0.059(3)	O	U_{11}	0.039(10)
		U_{22}	0.0170(14)		U_{22}	0.025(1)		U_{22}	0.018(8)
		U_{33}	0.1567(27)		U_{33}	0.072(1)		U_{33}	0.087(11)
	O2	U_{11}	0.1010(11)		U_{12}	−0.019(1)		U_{12}	0.018(8)
		U_{22}	0.1010(11)		U_{23}	0.011(2)		U_{23}	−0.012(3)
		U_{33}	0.0181(17)		U_{13}	0.016(2)		U_{13}	−0.012(3)
U_{eq} (Å^2)	O1	U_{eq}	0.0879(9)	O	U_{eq}	0.052(2)	O	U_{eq}	0.028(9)
	O2	U_{eq}	0.0734(8)						

Table 2.4 Comparison of equivalent isotropic atomic displacement parameters of $CsBi_2Ti_2NbO_{10-\delta}$ with those of layered perovskites with Bi–Bi–Ti bottleneck triangles. The U_{eq}(O1) and U_{eq}(O2) of $CsBi_2Ti_2NbO_{10-\delta}$ at 973 K are much higher than equivalent isotropic atomic displacement parameters of other oxides

Composition		$CsBi_2Ti_2NbO_{10-\delta}$				$BaBi_4Ti_4O_{15}$				$Bi_4Ti_3O_{12}$				$Bi_2Sr_2TiNb_2O_{12}$	
Temperature		RT		973 K		RT		1000 K		RT		923 K		RT	
Space group		*Ima*2		*P*4/*mmm*		*A*2$_1$*am*		*I*4/*mmm*		*Aba*2		*Aba*2		*B*2*cb*	
U_{eq} (Å^2)	Oxide-ion conducting layer	O1	0.0410(4)	O1	0.0879(9)	O2	0.023(1)	O1	0.032(2)	O1	0.032(2)	O1	0.059(2)	O1	0.0106(5)
		O2	0.0299(3)	O2	0.0734(8)	O3	0.023(1)	O2	0.023(1)	O3	0.014(1)	O3	0.049(1)	O3	0.0228(5)
		O3	0.0296(4)			O4	0.014(1)	O3	0.023(1)	O5	0.023(1)	O5	0.038(2)	O4	0.0113(5)
						O7	0.020(2)	O4	0.014(1)	O6	0.034(2)	O6	0.036(2)	O5	0.009(1)
						O8	0.013(1)								
	other site	O4	0.0085(3)	O3	0.0441(5)	O1	0.032(2)	O5	0.018(1)	O2	0.007(1)	O2	0.030(1)	O2	0.0046(5)
		O5	0.0089(3)	O4	0.0382(5)	O5	0.018(1)	O6	0.0094(7)	O4	0.010(1)	O4	0.039(1)	O6	0.007(1)
						O6	0.0094(7)								

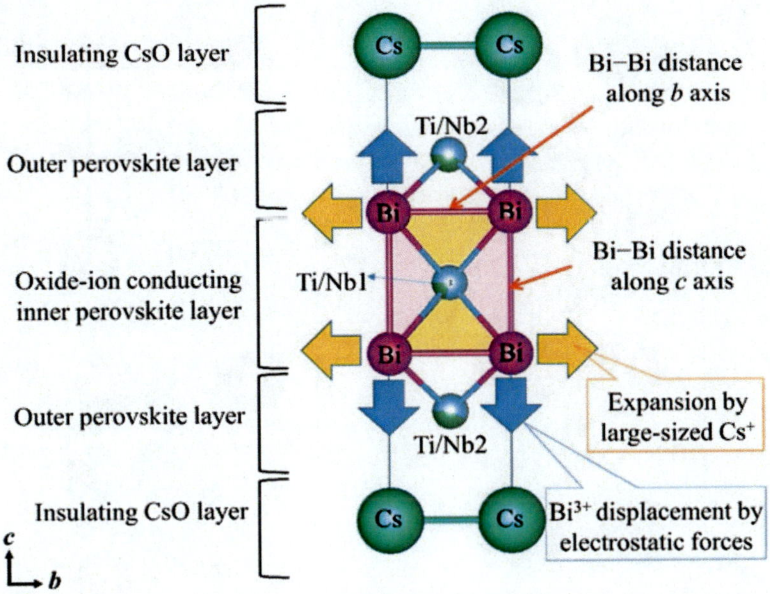

Fig. 2.37 New concept of enlarged bottlenecks for oxide-ion migration created by the large size of Cs^+ and Bi^{3+} displacement in Dion–Jacobson-type Cs/Bi-cation-ordered $CsBi_2Ti_2NbO_{10-\delta}$ ($-1/2 \leq x, y, z \leq 1/2$). Yellow and pink Bi–Bi–Ti triangles stand for the areas of bottlenecks for O1–O2 and O1–1 oxide-ion migration, respectively. Oxygen atoms are omitted for simplicity

was + 0.129(2) Å. The Bi^{3+} displacement in $CsBi_2Ti_2NbO_{10-\delta}$ can be explained by the electrostatic forces as described below. The electrostatic force F between two points charges is expressed by the Coulomb's law:

$$F = \frac{q_1 q_2}{4\pi \varepsilon_0 r^2}$$

where r is the distance between these two points charges q_1 and q_2, ε_0 is the permittivity of free space. To estimate the electrostatic force on a Bi^{3+} cation around $z = 0.13$, F_1, we consider the Coulomb's forces between the Bi^{3+} cation and four layers of $k = \pm 2$ and ± 1 (Fig. 2.39). F_1 is calculated as follows

$$F_1 = \frac{q_1 q_2}{4_0 r^2} = \frac{3e}{4\pi_0} \times \left[-\frac{2.30e}{r_2^2} - \frac{-1.89}{(2r_2)^2} + \frac{2.09e}{r_1^2} + \frac{1.05}{(2r_1)^2} \right]$$
$$\approx \frac{3e}{4\pi_0} \times \left[-\frac{2.30e}{r_1^2} - \frac{-1.89e}{(2r_1)^2} + \frac{2.09e}{r_1^2} + \frac{1.05e}{(2r_1)^2} \right]$$
$$\frac{3e}{4\pi_0} \times \frac{0.53e}{r_1^2} > 0$$

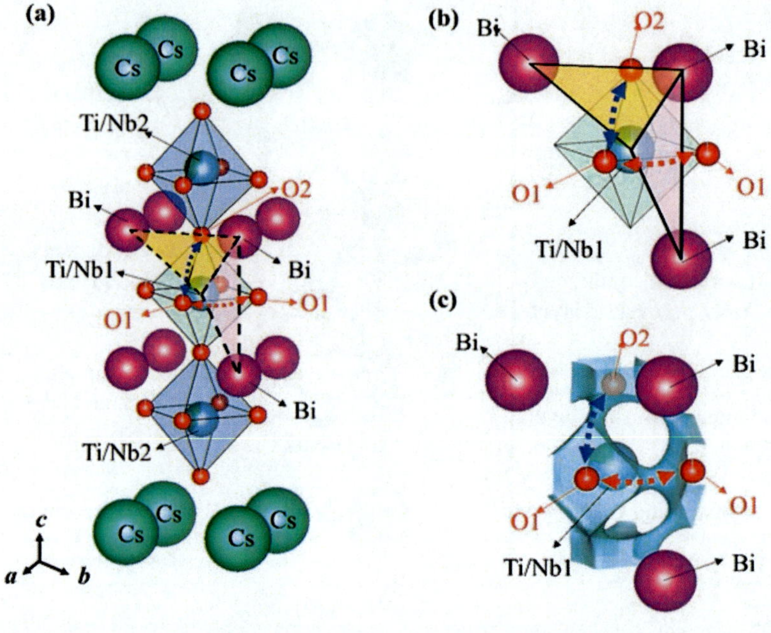

Fig. 2.38 Refined structure and Bi−Bi−Ti/Nb1 triangle bottleneck of $CsBi_2Ti_2NbO_{9.80(2)}$ at 973 K ($-1/2 \leq x, y, z \leq 1/2$), showing the new concept of enlarged bottlenecks for oxide-ion migration. A part of the structure **c** and corresponding BVEL at 0.6 eV **d** showing the O1−O2 and O1−O1 oxide-ion paths. Blue and red dotted lines with arrows denote the possible O1−O2 and O1−O1 diffusion pathways of oxide ions, respectively

where the interatomic distance r_2 is assumed to be r_1. Thus, the Bi^{3+} cation around $z = 0.13$ can be displaced upward along the c-axis (arrows in Fig. 2.39). Similarly, the electrostatic force on a Bi^{3+} cation around $z = -0.13$, F_2 is

$$F_2 \approx -\frac{3e}{4\pi\varepsilon_0} \times \frac{0.53e}{r_1^2} < 0$$

Thus, the Bi^{3+} cation around $z = -0.13$ can be displaced downward along the c-axis. These results indicate that the Bi^{3+} cations in $CsBi_2Ti_2NbO_{10-\delta}$ can be displaced apart from the Ti/Nb1−O1 (central Ti/Nb−O) layer due to the electrostatic forces. In $n = 3$ Dion–Jacobson phases A' $[A_{n-1}B_nO_{3n+1}]$ (=A' $[A_2B_3O_{10}]$, the A cation can be displaced apart from the central B−O layer due to the electrostatic forces which suggests that the $n = 3$ Dion–Jacobson phases could exhibit oxide-ion conduction.

The Bi^{3+} displacement expands the Bi−Bi distance along the c-axis. The expansions of Bi−Bi distances along both the b (= a) and c-axes lead to large bottleneck sizes for O1−O2 (yellow areas) and O1−O1 (pink areas) oxide-ion diffusion. Because there was no report of oxide-ion conductors in DJ phases, it is impossible to compare $CsBi_2Ti_2NbO_{10-\delta}$ with other DJ materials. In our present comparison,

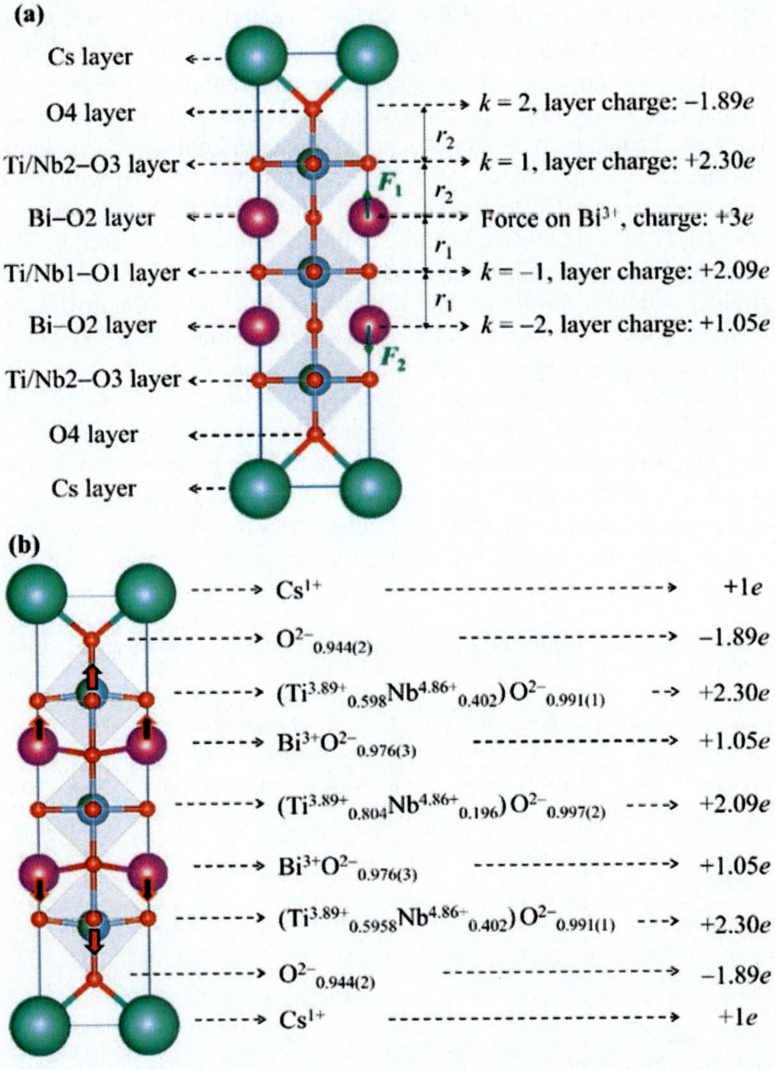

Fig. 2.39 Bi^{3+} displacement from the Bi−O2 layer due to electrostatic forces. (a) Hypothetical crystal structure of $CsBi_2Ti_2NbO_{10-\delta}$ without cation and anion displacements, consisting of Cs layers, O4 layers, Ti/Nb2−O3 layers, Bi−O2 layers, and Ti/Nb1−O1 layer ($-1/2 \leq x, y, z \leq 1/2$). (b) Refined crystal structure of $CsBi_2Ti_2NbO_{9.80(2)}$, which was obtained using neutron-diffraction data at 973 K ($-1/2 \leq x, y, z \leq 1/2$). r_1 and r_2 stand for the distances between Bi^{3+} and Ti/Nb1−O1 layer and Ti/Nb2−O3 layer. F_1 and F_2 are the electrostatic forces on the Bi^{3+} cations at around $z = 0.13$ and -0.13, respectively. A value in (b) stands for the charge of each layer

the oxide-ion conductivity of $CsBi_2Ti_2NbO_{10-\delta}$ is higher than those of the total conductivities, which means $\sigma_{ion}(CsBi_2Ti_2NbO_{10-\delta}) > \sigma_{total}(others) > \sigma_{ion}(others)$. Although they own different structures, all the materials own Bi–Bi–Ti bottleneck triangles for oxide-ion migration. Since the bottleneck of $CsBi_2Ti_2NbO_{10-\delta}$ is the Bi–Bi–Ti triangle, it is interesting to compare the bottlenecks and electrical conductivities of $CsBi_2Ti_2NbO_{10-\delta}$ with those of layered perovskites with Bi–Bi–Ti bottleneck triangles, $BaBi_4Ti_4O_{15}$, $Bi_4Ti_3O_{12}$, and $Bi_2Sr_2TiNb_2O_{12}$. The conductivity of $CsBi_2Ti_2NbO_{10-\delta}$ was much higher than those of these materials, which is attributable to the larger bottleneck sizes (Table 2.5) [23, 80, 81].

Aliovalent cations substitution and doping are effective to improve oxide-ion conductivity by increasing either oxygen vacancies or interstitial oxygen and by decreasing activation energy for oxide-ion conduction. In the present work, the $CsBi_{2-x}M_xTi_{2+y}Nb_{1-y}O_{10-x/2-y/2-\delta}$ (M = alkaline earth metal; $x = 0.1, 0.2, 0.3, 0.4, 0.5$ and 0.6; $y = 0.2, 0.1, -0.1$, and -0.2) compounds were investigated. In the series of $CsBi_{1.9}M_{0.1}Ti_2NbO_{9.95-\delta}$ (M = alkaline earth metal), the syntheses were succeeded without no impurities as shown in Fig. 2.40. However, as the generated oxygen vacancies by alkaline metals doping were slight, there was nearly no change in the total electrical conductivities (Fig. 2.41). Therefore, we increased the amount of doped

Table 2.5 Comparison of bottleneck size (critical radius) and total electrical conductivities of $CsBi_2Ti_2NbO_{10-\delta}$ with those of layered perovskites with Bi–Bi–Ti bottleneck triangles. The oxide-ion conductivity of $CsBi_2Ti_2NbO_{10-\delta}$ is approximated by its total electrical conductivity, which is higher than those of $BaBi_4Ti_4O_{15}$, $Bi_4Ti_3O_{12}$, and $Bi_2Sr_2TiNb_2O_{12}$. Thus, the oxide-ion conductivity of $CsBi_2Ti_2NbO_{10-\delta}$ is much higher than those of these materials

Composition	$CsBi_2Ti_2NbO_{10-\delta}$	$BaBi_4Ti_4O_{15}$	$Bi_4Ti_3O_{12}$	$Bi_2Sr_2TiNb_2O_{12}$
Structure type	Dion–Jacobson	Aurivillius	Aurivillius	Aurivillius
Oxide-ion conductivity at 1073 K (S cm^{-1})	1.3×10^{-2}	$< 4.0 \times 10^{-4}$	$< 1.0 \times 10^{-3}$	$< 1.3 \times 10^{-4}$
Critical radius (Å) ^{c}	1.0229(6)	0.9420(9)	0.9336(5)	0.9115(1)

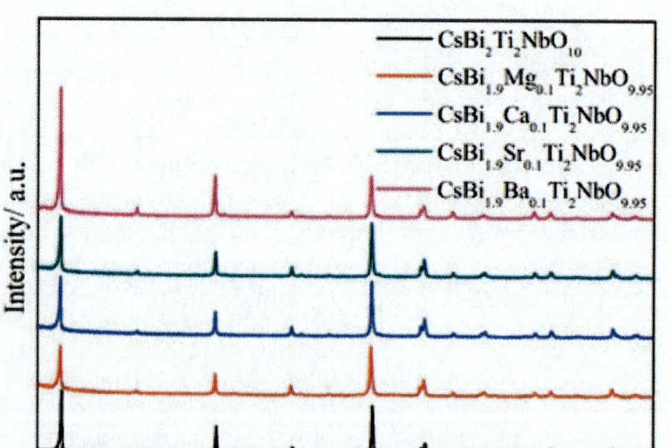

Fig. 2.40 X-ray powder diffraction patterns of $CsBi_{1.9}M_{0.1}Ti_2NbO_{9.95-\delta}$ (M = alkaline earth metal) and $CsBi_2Ti_2NbO_{10-\delta}$

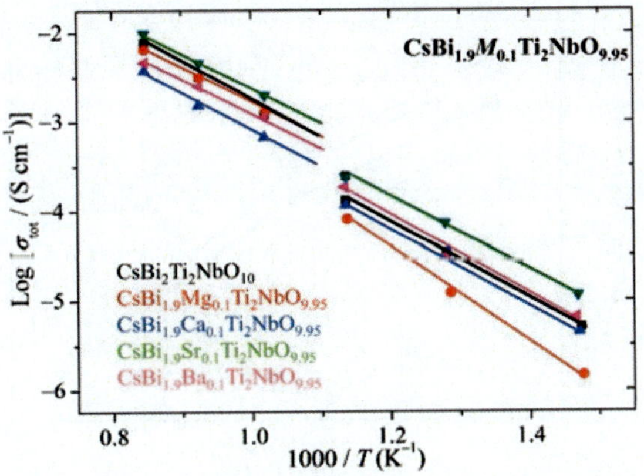

Fig. 2.41 Total DC electrical conductivities σ_{tot} of $CsBi_2Ti_2NbO_{10-\delta}$ and $CsBi_{1.9}M_{0.1}Ti_2NbO_{9.95-\delta}$.($M$ = alkaline earth metal) from 673 to 1173 K on heating in air, measured every 100 K and allowed 1 h of equilibration at each temperature step

level from 5 to 10% as shown in Fig. 2.42. The syntheses were successful except Mg-doped one. We attributed it to the small ionic size of Mg than Bi. We measured the conductivities of others and found the conductivities were improved (Fig. 2.43). As Ca-doped one showed the highest conductivities among $CsBi_{1.8}M_{0.2}Ti_2NbO_{9.8}$, we tried to increase the doped amount of Ca (Figs. 2.44 and 2.45). With the increasing amount of Ca, the conductivities increased as shown in Fig. 2.44. Unfortunately, a high level of impurities was observed in these Ca-doped compounds (Figs. 2.45 and 2.46). The impurity seems to be $Ca_{1.46}Ti_{1.38}Nb_{1.11}O_7$ (Fig. 2.46). Not only did doping on the *A* site (Bi), we also tried to do the doping on the *B* site (Ti and Nb). We changed the ratio of Ti and Nb in order to create some oxygen vacancies of

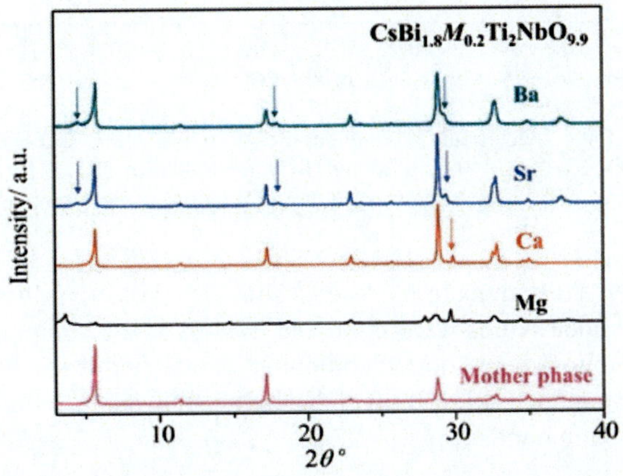

Fig. 2.42 X-ray powder diffraction patterns of $CsBi_{1.8}M_{0.2}Ti_2NbO_{9.9-\delta}$.($M$ = alkaline earth metal) and $CsBi_2Ti_2NbO_{10-\delta}$

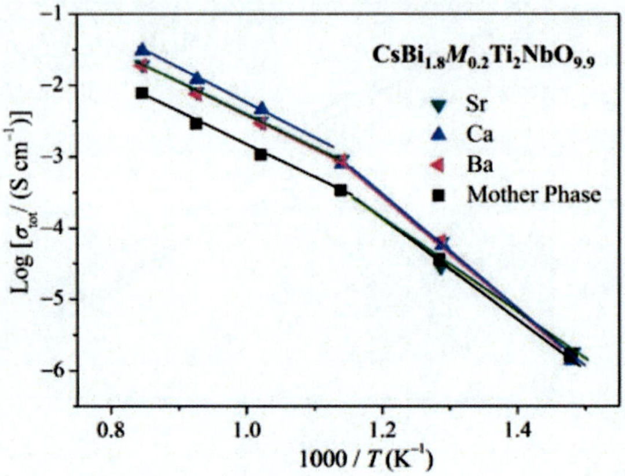

Fig. 2.43 Total DC electrical conductivities σ_{tot} of $CsBi_2Ti_2NbO_{10-\delta}$ and $CsBi_{1.8}M_{0.2}Ti_2NbO_{9.9-\delta}$.(alkaline earth metal) from 673 to 1173 K on cooling in air, measured every 100 K and allowed 1 h of equilibration at each temperature step

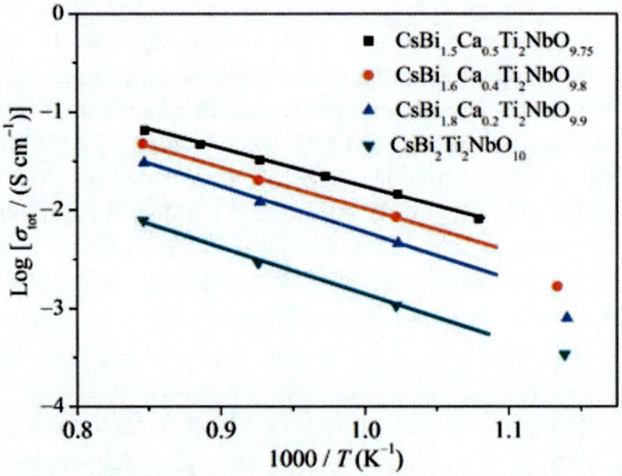

Fig. 2.44 Total DC electrical conductivities σ_{tot} of $CsBi_2Ti_2NbO_{10-\delta}$ and $CsBi_{2\text{-}x}M_{0.2}Ti_2NbO_{10-x/2}$.($x$ = 0.2, 0.4, and 0.5) from 873 to 1173 K on heating in air, measured every 100 K or 50 K and allowed 1 h of equilibration at each temperature step

interstitial oxygen. Although there existed a slight level of impurities (Fig. 2.47), the total electrical conductivities were improved in both of the compounds (Fig. 2.48). Furthermore, we chose some doped compounds to carry out the $P(O_2)$ dependents of total electrical conductivities. Both of these data showed the oxide-ion conduction in the measured compounds (Figs. 2.48, 2.49, 2.50, 2.51,2.52 and Table 2.6).

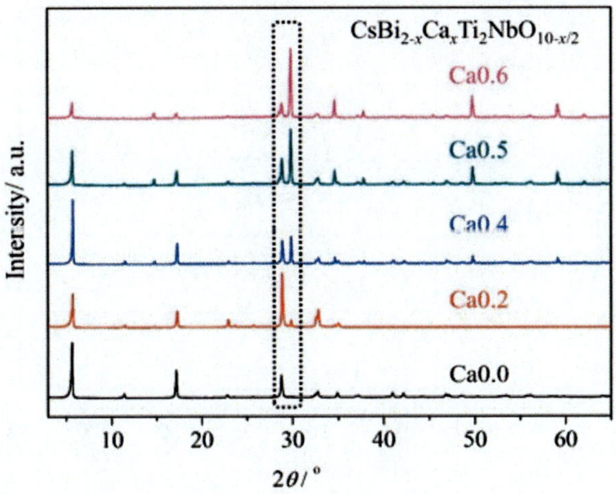

Fig. 2.45 X-ray powder diffraction patterns of $CsBi_{2-x}M_xTi_2NbO_{10-x/2-\delta}$. ($x = 0.2$, 0.4, 0.5, and 0.6) and $CsBi_2Ti_2NbO_{10-\delta}$

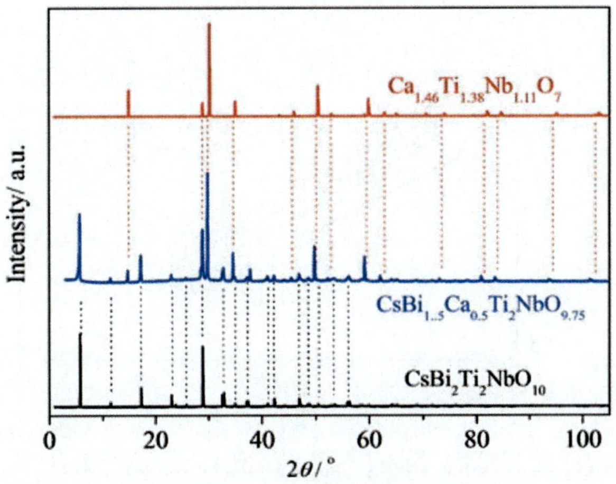

Fig. 2.46 X-ray powder diffraction patterns of $CsBi_{1.5}Ca_{0.5}Ti_2NbO_{9.75}$, $Ca_{1.46}Ti_{1.38}Nb_{1.11}O_7$, and $CsBi_2Ti_2NbO_{10-\delta}$

At last, the thermal expansion coefficients (TECs) of $CsBi_2Ti_2NbO_{10-\delta}$ in air, which were estimated using the lattice parameters were refined by the Rietveld analyses of the synchrotron X-ray diffraction data (Table 2.6). Thermal expansion anomalies were observed around the o-t phase transition temperature (Fig. 2.53). the TECs were close to that of 8 mol% Y_2O_3–ZrO_2 (8YSZ, 10.5×10^{-6} K^{-1}).

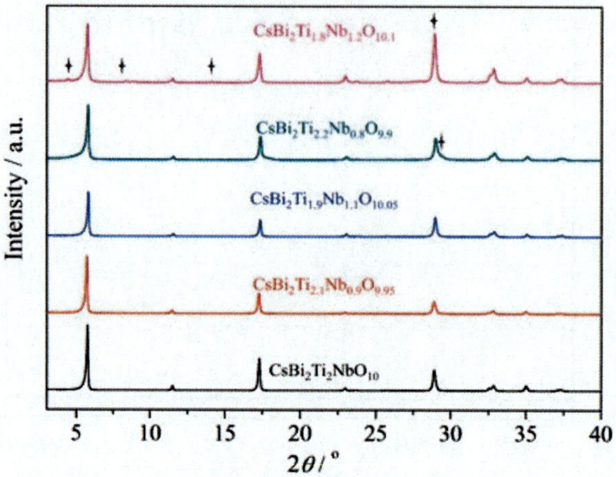

Fig. 2.47 X-ray powder diffraction patterns of $CsBi_2Ti_{2-x}Nb_xO_{10-x/2}$ (x = –0.2, –0.1, 0.1, and 0.2) and $CsBi_2Ti_2NbO_{10-\delta}$

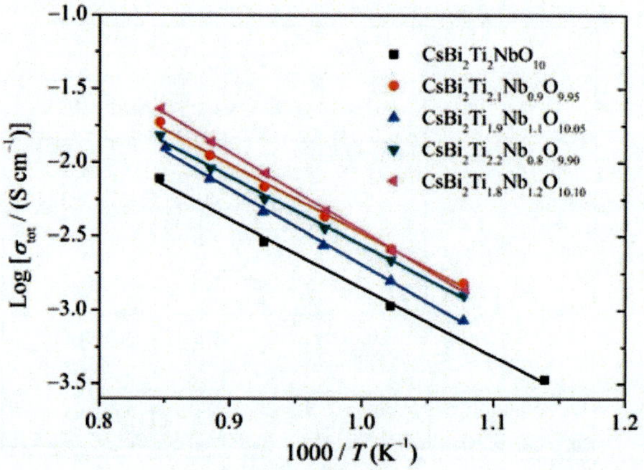

Fig. 2.48 Total DC electrical conductivities σ_{tot} of $CsBi_2Ti_2NbO_{10-\delta}$ and $CsBi_2Ti_{2-x}Nb_xO_{10-x/2}$ (x = –0.2, –0.1, 0.1, and 0.2) from 673 to 1173 K on heating in air, measured every 100 K or 50 K and allowed 1 h of equilibration at each temperature step

2.4 Conclusion

In summary, we have found the first example of the Dion–Jacobson-type layered perovskite of oxide-ion conductor $CsBi_2Ti_2NbO_{10-\delta}$. The Cs/Bi-cation-ordered $CsBi_2Ti_2NbO_{10-\delta}$ shows good bulk ionic conductivity of 8.9×10^{-2} S cm^{-1} at

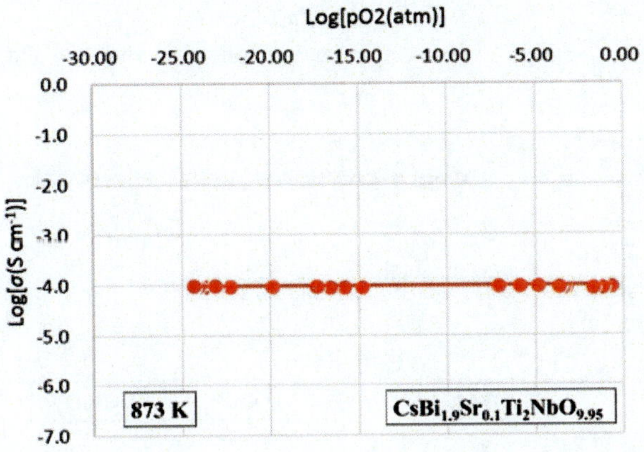

Fig. 2.49 Partial oxygen pressure $P(O_2)$ dependence of total DC electrical conductivities σ_{tot} of $CsBi_{1.9}Sr_{0.1}Ti_2NbO_{9.95-\delta}$ at 873 K

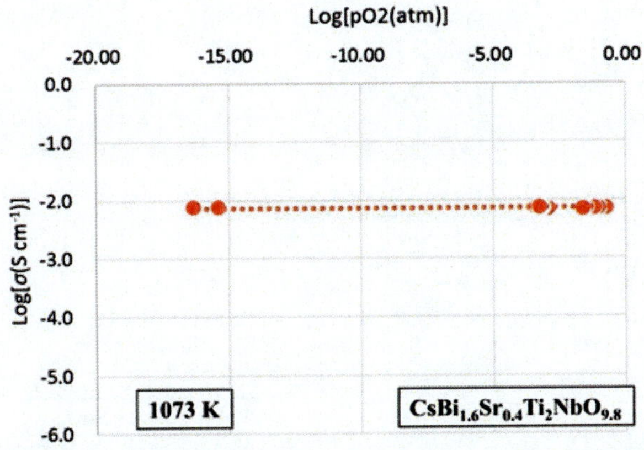

Fig. 2.50 Partial oxygen pressure $P(O_2)$ dependence of total DC electrical conductivities σ_{tot} of $CsBi_{1.6}Sr_{0.4}Ti_2NbO_{9.8-\delta}$ at 1073 K

1073 K and high phase stability at high-temperature ranges and at wide $P(O_2)$ ranges. The ionic conductivities showed an abrupt increase around 673–873 K on heating, which may be ascribed to the increase of oxygen vacancies and also the orthorhombic-to-tetragonal phase transition. The novel oxide-ion conductivities of the $CsBi_2Ti_2NbO_{10-\delta}$ are attributable to (i) the large anisotropic thermal motions of oxygen atoms in the inner oxide-conducting perovskite layer, (ii) 2D O1−O2 and O1−O1 oxide-ion diffusion, (iii) existence of oxygen vacancies, and (iv) large bottleneck sizes. Here, we proposed a new concept, which is the large bottlenecks for

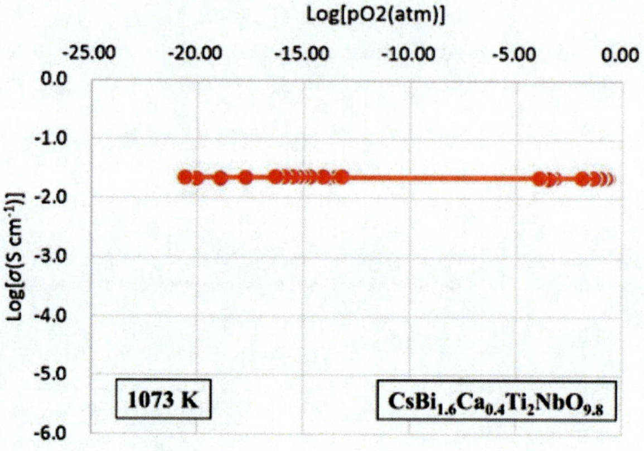

Fig. 2.51 Partial oxygen pressure $P(O_2)$ dependence of total DC electrical conductivities σ_{tot} of $CsBi_{1.6}Ca_{0.4}Ti_2NbO_{9.8-\delta}$ at 1073 K

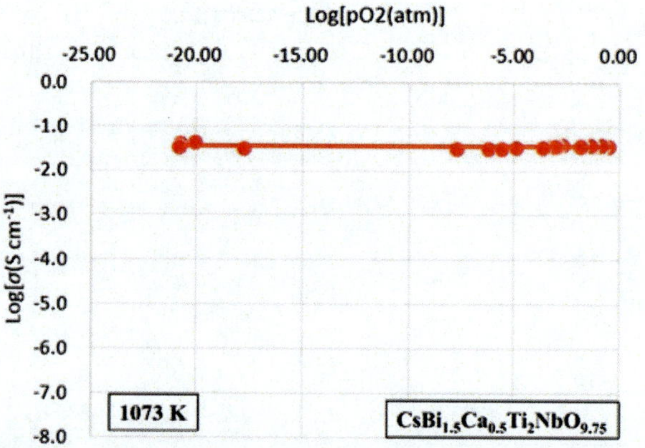

Fig. 2.52 Partial oxygen pressure $P(O_2)$ dependence of total DC electrical conductivities σ_{tot} of $CsBi_{1.5}Ca_{0.5}Ti_2NbO_{9.75-\delta}$ at 1073 K

oxide-ion migration by the large size of Cs^+ and Bi^{3+} displacement. The oxide-ion conductivity is expected to be improved by doping and/or modification of the chemical composition of the basic material of $CsBi_2Ti_2NbO_{10-\delta}$. The present finding of high oxide-ion conductivities in the new structure family Dion–Jacobson-type $CsBi_2Ti_2NbO_{10-\delta}$ and the concept would open new windows in the design and discovery of the new and novel oxide-ion conductors based on the Dion–Jacobson phases.

Table 2.6 Average thermal expansion coefficients of $CsBi_2Ti_2NbO_{10-\delta}$ in air, which were estimated using the lattice parameters refined by the Rietveld analyses of the synchrotron X-ray diffraction data

Compositions	$CsBi_2Ti_2NbO_{10-\delta}$		
Temperature range (K)	298–673	298–873	298–1073
$\overline{\alpha}^o_b$ or $\overline{\alpha}^t_b$($\times$ 10^{-6} K^{-1})	12.79	15.57	14.14
$\overline{\alpha}^o_c$ or $\overline{\alpha}^t_a$ ($\times$ 10^{-6} K^{-1})	5.93	0.81	3.16
$\overline{\alpha}^o_a$ or $\overline{\alpha}^t_c$($\times$ 10^{-6} K^{-1})	17.36	19.27	18.32
$\overline{\alpha}_L$($\times$ 10^{-6} K^{-1})	12.02	11.87	11.86

The average TECs in the temperature range between T and T_0 are defined as follows:

$\overline{\alpha}^o_b = (b^o(T) - b^o(T_0))/b^o(T_0) / (T - T_0), \overline{\alpha}^t_b = (\sqrt{2}a^t(T) - b^o(T_0)) / b^o(T_0) / (T - T_0),$

$\overline{\alpha}^o_c = (c^o(T) - c^o(T_0))/c^o(T_0) / (T - T_0), \overline{\alpha}^t_a = (\sqrt{2}\,a^t(T) - c^o(T_0)) / c^o(T_0) / (T - T_0),$

$\overline{\alpha}^o_a = (a^o(T) - a^o(T_0))/a^o(T_0) / (T - T_0), \overline{\alpha}^t_c = (\sqrt{2}\,c^t(T) - a^o(T_0)) / a^o(T_0) / (T - T_0),$

$\overline{\alpha}_L = (v(T) - v(T_0)) / v(T_0) / (T - T_0)$ where $v(T)[a^o(T)\,b^o(T)\,c^o(T)]^{1/3}$

or $v(T)[4a^t(T)\,a^t(T)\,c^t(T)]^{1/3}$

where $T_0 = 298$ K and the superscripts o and t denote the orthorhombic and tetragonal, respectively.

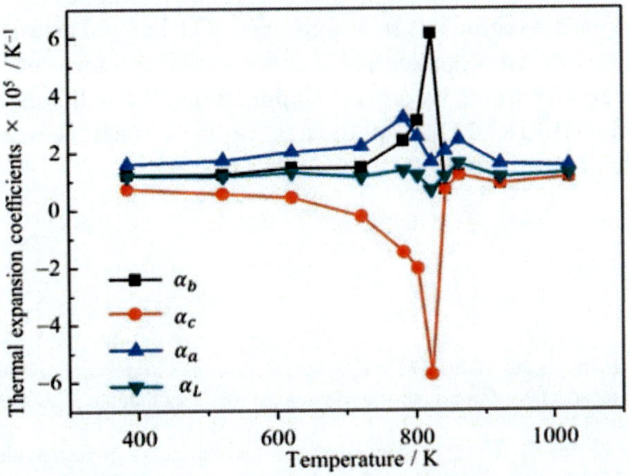

Fig. 2.53 Thermal expansion coefficients (TECs) of $CsBi_2Ti_2NbO_{10-\delta}$ in air, which were estimated using the lattice parameters refined by the Rietveld analyses of the synchrotron X-ray diffraction data (Fig. 1a). Thermal expansion anomalies were observed around the o-t phase transition temperature. TECs are defined as follows:

$\alpha_b = (b(Ti + 1)b(Ti))/b(Ti)/(Ti + 1 - Ti)$ where $b(Ti) = b^o(Ti)$ *or* $\sqrt{2}\,a^t(Ti)$

$\alpha_c = (c(T_{i+1}) - c(T_i))/c(T_i)/(T_{i+}1 - T_i)$ where $c(T_i) = c^\circ(T_i)$ or $\sqrt{2} \cdot a^t(T_i)$

$\alpha_a = (a(T_i + 1)a(T_i))/a(T_i)/(T_i + 1 - T_i)$ where $a(T_i) = a^o(T_i)$ or $2\,c^t(T_i)$

$\alpha_L = (v(T_{i+1}) - v(T_i))/v(T_i)/(T_{i+1} - T_i)$ where $v(T_i)\ [a^\circ(T_i) \cdot b^\circ(T_i) \cdot c^\circ(T_i)]^{1/3}$

or $v(T_i) = [4 \cdot a^t(T_i)a^t(T_i) \cdot c^t(T_i)]^{1/3}$

$T_1 = 297$ K, $T_2 = 473$ K, $T_3 = 573$ K, $T_4 = 673$ K, $T_5 = 773$ K, $T_6 = 793$ K, $T_7 = 813$ K (orthorhombic phase). $T_8 = 833$ K, $T_9 = 853$ K, $T_{10} = 873$ K, $T_{11} = 973$ K, $T_{12} = 1073$ K (tetragonal phase)

2.5 Outlook

The novel oxide-ion conductivities of the Dion–Jacobson-type $CsBi_2Ti_2NbO_{10-\delta}$, belong to a new structural family, can open new windows to the design of novel oxide-ion conductors. However, there also exist some problems that need to be overcome in the oxide-ion conductor of $CsBi_2Ti_2NbO_{10-\delta}$ as follows.

Dion–Jacobson phase $CsBi_2Ti_2NbO_{10-\delta}$ exists a phase transition around 873 K, which just happened at the operating temperature of SOFCs. The phase transition may lead to some problems during the practice application, such as mismatching of the coefficient of thermal expansion with an electrode and low conductivity at low temperature. This problem could be solved using some modifications. We have confirmed that there was no phase transition in the similar Dion–Jacobson structure of $CsLa_2Ti_2NbO_{10}$, which will be discussed in Chap. 3. There is an urgent need to discuss the possibility of $CsBi_{1-x}La_xTi_2NbO_{10,}$ which may keep the high oxide-ion conductivity without a phase transition or decrease the phase transition temperature. This work will be done by our group in the future.

The oxide-ion conductivity below 873 K was insufficient as an electrolyte of SOFCs that need an oxide-ion conductivity as high as ~10^{-2} S cm^{-1} at the operating temperature. Although we have studied some doping compounds, the conductivities at low-temperature regions were unimproved. Further and detailed research was asked. In addition, electron conduction occurred at a high level of reducing atmosphere (H_2) which also limits its broader applications. We will continue to improve the properties of $CsBi_2Ti_2NbO_{10-\delta}$ in the near future to meet the need for a practice oxide-ion conductor.

References

1. Malavasi L, Fisher CAJ, Islam MS (2010) Oxide-ion and proton conducting electrolyte materials for clean energy applications: Structural and mechanistic features. Chem Soc Rev 39(11):4370–4387
2. Boivin JC, Mairesse G (1998) Recent material developments in fast oxide-ion conductors. Chem Mater 10(10):2870–2888
3. Goodenough JB (2000) Oxide-ion conductors by design. Nature 404(6780):821–823
4. Kharton VV, Yaremchenko AA, Naumovich EN, Marques FMB (1999) Research on the electrochemistry of oxygen ion conductors in the former Soviet Union. J Solid State Electrochem 3:303–326
5. Skinner SJ, Kilner JA (2003) Ion conductors. Rev Lit Arts Am 6(3):30–37
6. Ishihara T, Matsuda H, Takita Y (1994) Doped LaGaO3 perovskite type oxide as a new oxide ionic conductor. J Am Chem Soc 116(9):3801–3803
7. Ishihara T, Kilner JA, Honda M, Takita Y (1997) Oxygen surface exchange and diffusion in the new perovskite oxide-ion conductor LaGaO3. J Am Chem Soc 119(11):2747–2748
8. Malavasi L, Kim HJ, Billinge SJL, Proffen T, Tealdi C, Flor G (2007) Nature of the monoclinic to cubic phase transition in the fast oxygen ion conductor La2Mo2O9 (LAMOX). J Am Chem Soc 129(21):6903–6907

9. Manthiram A, Kim JH, Kim YN, Lee KT (2011) Crystal chemistry and properties of mixed ionic-electronic conductors. J Electroceram 27(2):93–107
10. Lacorre P, Goutenoire Ë, Bohnke O, Retoux R, Laligant Y (2000) Designing fast oxide-ion conductors based on La2Mo2O9. Nature 404:856
11. Yashima M (2009) Diffusion pathway of mobile ions and crystal structure of ionic and mixed conductors–a brief review. J Ceram Soc Japan 117(1370):1055–1059
12. Yashima M (2015) Invited review: some recent developments in the atomic-scale characterization of structural and transport properties of ceria-based catalysts and ionic conductors. Catal Today 253:3–19
13. FujiiI K, Yashima M (2018) Discovery and development of BaNdInO4—a brief review. J Ceram Soc Japan 126(10):852–859
14. Yashima M, Sirikanda N, Ishihara T (2010) Crystal structure, diffusion path, and oxygen permeability of a Pr 2NiO4-based mixed conductor (Pr0.9la 0.1)2(Ni0.74Cu0.21Ga 0.05)O4+δ. J Am Chem Soc 132(7):2385–2392
15. Yashima M et al (2008) Structural disorder and diffusional pathway of oxide ions in a doped Pr2NiO4-based mixed conductor. J Am Chem Soc 130(9):2762–2763
16. Fujii K et al (2014) New perovskite-related structure family of oxide-ion conducting materials NdBaInO4. Chem Mater 26(8):2488–2491
17. Fujimoto A, Yashima M, Fujii K, Hester JR (2017) New oxide-ion conductor SrYbInO4 with partially cation-disordered CaFe2O4-type structure. J Phys Chem C 121(39):21272–21280
18. Abraham F, Boivin JC, Mairesse G, Nowogrocki G (1990) The bimevox series: a new family of high performances oxide-ion conductors. Solid State Ionics 40–41(2):934–937
19. Iharada T, Hammouche A, Fouletier J, Kleitz M, Boivin JC, Mairesse G (1991) Electrochemical characterization of BIMEVOX oxide-ion conductors. Solid State Ionics 48(3–4):257–265
20. Pirovano C, Steil MC, Capoen E, Nowogrocki G, Vannier RN (2005) Impedance study of the microstructure dependence of the electrical properties of BIMEVOXes. Solid State Ionics 176(25–28):2079–2083
21. Lee CK, Westb AR (1996) Thermal behaviour and polymorphism conductors including the new materials: Mg, B IONICS of BIMEVOX oxide ion Bi,V,O, 1 : M; M = La, Y. Solid State Ionics 2738(96)
22. Kendall KR, Navas C, Thomas JK, Zur Loye HC (1996) Recent developments in oxide-ion conductors: aurivillius phases. Chem Mater 8(3):642–649
23. Thomas JK, Anderson ME, Krause WE, Zur Loye H (1993) Oxygen ion conductivity in a new class of layered bismuth oxide compounds. MRS Online Proc Libr Arch 293:295–300
24. Porob DG, Guru Row TN (2000) A novel oxide-ion conductor in a doped Bi2O3-V2O5 system: Ab initio structure of a new polymorph of NaBi3V2O10 via powder X-ray diffraction. Chem Mater 12(12):3658–3661
25. Amow G, Skinner SJ (2006) Recent developments in Ruddlesden-Popper nickelate systems for solid oxide fuel cell cathodes. J Solid State Electrochem 10(8):538–546
26. Patrakeev MV, Leonidov IA, Kozhevnikov VL, Kharton VV (2004) Ion-electron transport in strontium ferrites: relationships with structural features and stability. Solid State Sci 6(9):907–913
27. Yamada A et al (2008) Ruddlesden-popper-type epitaxial film as oxygen electrode for solid-oxide fuel cells. Adv Mater 20(21):4124–4128
28. Yang G, Su C, Ran R, Tade MO, Shao Z (2014) Advanced symmetric solid oxide fuel cell with an infiltrated K 2NiF4-Type La2NiO4 electrode. Energy Fuels 28(1):356–362
29. Ishihara T, Nakashima K, Okada S, Enoki M, Matsumoto H (2008) Defect chemistry and oxygen permeation property of Pr2Ni0.75Cu0.25O4 oxide doped with Ga. Solid State Ionics 179(27–32):1367–1371
30. Sengodan S et al (2015) Layered oxygen-deficient double perovskite as an efficient and stable anode for direct hydrocarbon solid oxide fuel cells. Nat Mater 14(2):205–209
31. Ali R, Yashima M, Izumi F (2007) Diffusion path of oxide ions in an oxide-ion conductor la 0.64(Ti0.92Nb0.08)O2.99 with a double perovskite-type structure. Chem Mater 19(13):3260–3264

32. Burriel M et al (2012) Anisotropic oxygen ion diffusion in layered $PrBaCo_2O_{5+\delta}$. Chem Mater 24(3):613–621
33. Goodenough JB, Ruiz-Diaz JE, Zhen YS (1990) Oxide-ion conduction in Ba2In2O5 and Ba3In2MO8 (M=Ce, Hf, or Zr). Solid State Ionics 44(1–2):21–31
34. Auckett JE et al (2013) Combined experimental and computational study of oxide ion conduction dynamics in Sr2Fe2O5 brownmillerite. Chem Mater 25(15):3080–3087
35. Fop S, McCombie KS, Wildman EJ, Skakle JMS, Mclaughlin AC (2019) Hexagonal perovskite derivatives: a new direction in the design of oxide ion conducting materials. Chem Commun 55(15):2127–2137
36. Fop S et al (2016) Oxide ion conductivity in the hexagonal perovskite derivative Ba3MoNbO8.5. J Am Chem Soc 138(51):16764–16769
37. Yashima M, et al (2019) Direct evidence for two-dimensional oxide-ion diffusion in the hexagonal perovskite-related oxide $Ba3MoNbO_{8.5-\delta}$. J. Mater Chem A
38. Fujii K, Shiraiwa M, Esaki Y, Yashima M, Kim SJ, Lee S (2015) Improved oxide-ion conductivity of NdBaInO4 by Sr doping. J Mater Chem A 3(22):11985–11990
39. Yang X, Liu S, Lu F, Xu J, Kuang X (2016) Acceptor doping and oxygen vacancy migration in layered perovskite NdBaInO4-based mixed conductors. J Phys Chem C 120(12):6416–6426
40. Shiraiwa M, Fujii K, Esaki Y, Kim SJ, Lee S, Yashima M (2017) Crystal structure and oxide-ion conductivity of Ba 1+ x Nd 1−x InO 4−x /2. J Electrochem Soc 164(13):F1392–F1399
41. Dion M, Ganne M, Tournoux M (1981) Nouvelles familles de phases MIMII2Nb3O10 a feuillets 'perovskites.' Mater Res Bull 16(11):1429–1435
42. Jacobson AJ, Johnson JW, Lewandowski JT (1985) Interlayer chemistry between thick transition-metal oxide layers: synthesis and intercalation reactions of K[Ca2Nan-3NbnO3+1] ($3 \leq n \leq 7$). Inorg Chem 24(23):3727–3729
43. Mccabe EE et al (2015) Proper ferroelectricity in the dion-jacobson material CsBi2Ti2NbO10: experiment and theory. Chem Mater 27(24):8298–8309
44. Goff RJ, Keeble D, Thomas PA, Ritter C, Morrison FD, Lightfoot P (2009) Leakage and proton conductivity in the predicted ferroelectric CsBiNb 2O7. Chem Mater 21(7):1296–1302
45. Draskovic TI, Wang T, Henderson CN, Mallouk TE (2014) Protonic and electronic conductivity of the layered perovskite oxides HCa2Nb3O10 and Ca4Nb 6O19. Int J Hydrogen Energy 39(9):4576–4580
46. Kim Y, Kim SJ (2015) Ionic conductivity of Dion–Jacobson type oxide LiLaTa2O7and oxynitride LiLaTa2O6.15N0.57measured by impedance spectroscopy. Ceram Int 41(2):3318–3323
47. Choy JH, Kim JY, Kim SJ, Sohn JS, Hee Han OC (2001) New Dion-Jacobsen-type layered perovskite oxyfluorides, ASrNb2O6F (A = Li, Na, and Rb). Chem Mater 13(3):906–912
48. Thangadurai V, Weppner W (2010) ChemInform abstract: determination of the sodium ion transference number of the dion–jacobson-type layered perovskite NaCa2Nb3O10 using ac impedance and dc methods. ChemInform 33(24)
49. Thangadurai V, Weppner W (2004) Mixed potential protonic-electronic conductivity in the Dion–Jacobson-type layered perovskites in hydrogen-containing atmosphere and their application in ammonia sensors. Solid State Ionics 174(1–4):175–183
50. Inoue R, Fujii K, Shiraiwa M, Niwa E, Yashima M (2018) A new structure family of oxide-ion conductors Ca0.8Y2.4Sn0.8O6 discovered by a combined technique of the bond-valence method and experiments. Dalt Trans 47(22):7515–7521
51. Niwa E, Yashima M (2018) Discovery of oxide-ion conductors with a new crystal structure, BaSc 2−x A x Si 3 O 10−x/2 (A = Mg, Ca) by screening Sc-containing oxides through the bond-valence method and experiments. ACS Appl Energy Mater 1(8):4009–4015
52. Yasui Y, Niwa E, Matsui M, Fujii K, Yashima M (2019) Discovery of a rare-earth-free oxide-ion conductor Ca3Ga4O9 by screening through bond valence-based energy calculations, synthesis, and characterization of structural and transport properties. Inorg Chem 58(14):9460–9468
53. Adams S (2001) Relationship between bond valence and bond softness of alkali halides and chalcogenides. Acta Crystallogr Sect B Struct Sci 57(3):278–287
54. Nakamura K, Fujii K, Niwa E, Yashima M (2018) Crystal structure and electrical conductivity of BaR2ZnO5 (R = Sm, Gd, Dy, Ho, and Er)—a new structure family of oxide-ion conductors. J Ceram Soc Japan

55. Avdeev M, Sale M, Adams S, Rao RP (2012) Screening of the alkali-metal ion containing materials from the inorganic crystal structure database (ICSD) for high ionic conductivity pathways using the bond valence method. Solid State Ionics 225:43–46
56. Nishibori E et al (2001) The large Debye-Scherrer camera installed at SPring-8 BL02B2 for charge density studies. J Phys Chem Solids 62(12):2095–2098
57. Kawaguchi S, et al (2017) High-throughput powder diffraction measurement system consisting of multiple MYTHEN detectors at beamline BL02B2 of SPring-8. Rev Sci Instrum 88(8)
58. Rouse KD, Cooper MJ, York EJ, Chakera A (1970) Absorption corrections for neutron diffraction. Acta Crystallogr Sect A 26(6):682–691
59. Torii S, et al (2011) Super high resolution powder diffractometer at J-PARC. J. Phys. Soc. Japan 80(Suppl. B):3–6
60. Yashima M, Ali R (2009) Structural phase transition and octahedral tilting in the calcium titanate perovskite CaTiO3. Solid State Ionics 180(2–3):120–126
61. Adams S (2006) From bond valence maps to energy landscapes for mobile ions in ion-conducting solids. Solid State Ionics 177(19–25 SPEC. ISS): 1625–1630
62. Oishi R, et al (2009) Rietveld analysis software for J-PARC. Nucl Instruments Methods Phys Res Sect A Accel Spectrometers Detect Assoc Equip 600(1):94–96
63. Momma K, Izumi F (2011) VESTA 3 for three-dimensional visualization of crystal, volumetric and morphology data. J Appl Crystallogr 44(6):1272–1276
64. Momma K, Ikeda T, Belik AA, Izumi F (2013) Dysnomia, a computer program for maximum-entropy method (MEM) analysis and its performance in the MEM-based pattern fitting. Powder Diffr 28(3):184–193
65. Belkly A, Helderman M, Karen VL, Ulkch P (2002) New developments in the inorganic crystal structure database (ICSD): accessibility in support of materials research and design. Acta Crystallogr Sect B Struct Sci 58(3 PART 1): 364–369
66. Chen H, Wong LL, Adams S (2019) SoftBV—a software tool for screening the materials genome of inorganic fast ion conductors. Acta Crystallogr Sect B Struct Sci Cryst Eng Mater 75(1):18–33
67. Adams S (2000) Modelling ion conduction pathways by bond valence pseudopotential maps. Solid State Ionics 136–137:1351–1361
68. West A, Irvine J, Sinclair D (1990) Electroceramics: characterization by impedance spectroscopy. Adv Mater 2(3):132–138
69. Kwon OH, Choi GM (2006) Electrical conductivity of thick film YSZ. Solid State Ionics 177(35–36):3057–3062
70. Li M et al (2014) A family of oxide-ion conductors based on the ferroelectric perovskite Na0.5Bi0.5TiO3. Nat Mater 13(1):31–35
71. Huang K, Feng M, Goodenough JB (1998) Synthesis and electrical properties of dense Ce0.9Gd0.1O1.95 ceramics. J Am Chem Soc 81(2):357–362
72. Kuang X et al (2008) Interstitial oxide ion conductivity in the layered tetrahedral network melilite structure. Nat Mater 7(6):498–504
73. Yashima M, Itoh M, Inaguma Y, Morii Y (2005) Crystal structure and diffusion path in the fast lithium-ion conductor La0.62Li0.16TiO3. J Am Chem Soc 127(10):3491–3495
74. Chen YC, Yashima M, Peña-Martínez J, Kilner JA (2013) Experimental visualization of the diffusional pathway of oxide ions in a layered perovskite-type cobaltite PrBaCo2O5+δ. Chem Mater 25(13):2638–2641
75. Yashima M, Ishimura D (2003) Crystal structure and disorder of the fast oxide-ion conductor cubic Bi2O3. Chem Phys Lett 378(3–4):395–399
76. Yashima M, Kamioka T (2008) Neutron diffraction study of the perovskite-type lanthanum cobaltite. Solid State Ionics 178:1939–1943
77. Bhosale DR, Yusuf SM, Kumar A, Mukadam MD, Patil SI (2017) High oxide ion conductivity below 500 C in the garnets L ax Y3–x F e5 O12+δ. Phys Rev Mater 1(1):2–6
78. Slater PR, Irvine JTS, Ishihara T, Takita Y (1998) High-temperature powder neutron diffraction study of the oxide-ion conductor La0.9Sr0.1Ga0.8Mg0.2O2.85. J Solid State Chem 143(139):135–143

79. Yashima M, Nomura K, Kageyama H, Miyazaki Y, Chitose N, Adachi K (2003) Conduction path and disorder in the fast oxide-ion conductor (La0.8Sr0.2)(Ga0.8Mg0.15Co0.05)O2.8. Chem Phys Lett 380(3–4):391–396
80. Long C, Fan H, Ren W, Zhao J (2019) Double polarization hysteresis and dramatic influence of small compositional variations on the electrical properties in Bi4Ti3O12 ceramics. J Eur Ceram Soc 39(14):4103–4112
81. Shi J (2015) Crystal structure studies, electrical and magnetic properties of 2, 3, 4, 5-layer aurivillius oxides

Chapter 3
New Oxide-Ion Conductors of Dion–Jacobson-Type Layered Perovskites $CsLn_2Ti_2NbO_{10}$ (Ln = La, Pr, Nd, Sm, Gd, Ho, and Yb) and $CsLa_{2-x}M_xTi_{2+y}Nb_{1-y}O_{10-x/2-y/2}$ (M = Ca, Sr, and Ba; x = 0, 0.1; y = 0.1, 0, and –0.1)

Abstract As introduced in the Chap. 2, Dion–Jacobson-type layered perovskites have attracted much attention because of their novel physical and chemical properties such as ferroelectricity, photocatalyst, magnetism, and electrical conduction. The research in the caption presents a series of Dion–Jacobson phases of oxide-ion conductors. Dion–Jacobson phase $CsLa_2Ti_2NbO_{10-\delta}$ and its related compounds of $CsLn_2Ti_2NbO_{10}$ (Ln = La, Pr, Nd, Sm, Gd, Ho, and Yb) and $CsLa_{2-x}M_xTi_{2+y}Nb_{1-y}O_{10-x/2-y/2}$ (M = Ca, Sr, and Ba; x = 0, 0.1; y = 0.1, 0, and –0.1) were synthesized and some of the compositions' electrical conductivity and crystal structure were investigated from room temperature to high temperature. Unlike the $CsBi_2Ti_2NbO_{10-\delta}$ discussed in the Chap. 2, Rietveld's analysis of high-temperature neutron and synchrotron X-ray diffraction data revealed no phase transition from room temperature to 800 °C in the mother phase of $CsLa_2Ti_2NbO_{10-\delta}$, exhibiting good practical realizations and applications. Bond-valence-based energy landscapes of a test oxide ion show two-dimensional oxide-ion diffusion along the edges of an octahedron in the perovskite layers. The larger-sized Cs in $CsLa_2Ti_2NbO_{10-\delta}$ make the larger bottlenecks for oxide-ion migration, which leads to the lower activation energy for the oxide-ion conduction. The present finding of oxide-ion conductivities in the Dion–Jacobson phase $CsLa_2Ti_2NbO_{10-\delta}$ indicates that the wide possibility of Dion–Jacobson phase as oxide-ion conductors would open up new horizons in the discovery of other novel oxide-ion conductors, which based on the reported Dion–Jacobson phases.

Parts of this paper have been published in the following journal articles.

Zhang, W., Fujii, K., Ishiyama, T., Kandabashi, H., & Yashima, M. (2020). Dion–Jacobson-type oxide-ion conductor CsLa 2 Ti 2 NbO 10 − δ without phase transitions. *Journal of Materials Chemistry A*, *8*(47), 25,085–25,093.

W. Zhang, *Oxide-Ion Conduction in the Dion–Jacobson-Type Layered Perovskites*, Springer Theses, https://doi.org/10.1007/978-981-19-2247-3_3

3.1 Introduction

Oxide-ion (O^{2-}) conductors are inorganic materials exhibiting oxide-ion conduction. They have attracted much attention because of their extensive applications in energy conversion devices and environmental protection fields, for example, solid-oxide fuel cells (SOFCs), gas sensors, oxygen separation membranes, and oxygen-based catalysts. However, high oxide-ion conductivities have been reported only in a limited number of structural families as we discussed in Chap. 2. Since the few structural families limit the further development of chemistry and solid-state ionics, there is thereby an urgent need to discover new structural families of oxide-ion conductors to enhance the performance of the aforementioned electrochemical devices [1–10].

Some layered perovskites such as BIMEVOX, [11, 11] Aurivillius phase, [13, 14] Ruddlesden–Popper phase, [15–17] double perovskite, [18, 19] brownmillerite, [20, 21] hexagonal perovskite derivatives [22, 23], and others [24–26] have been reported to exhibit high oxide-ion conductivities. The Dion–Jacobson phase is also a layered perovskite with a general formula $A'[A_{n-1}B_nO_{3n+1}]$ [27, 28]. There have been numerous studies on the electrical properties such as proton, lithium-ion, sodium-ion, and mixed protonic–electronic conduction in Dion–Jacobson phases [29–31]. However, there have been no reports of oxide-ion conduction in the Dion–Jacobson phases. Thus, the discovery of Dion–Jacobson-type oxide-ion conductors would be important and challenging in the inorganic chemistry and materials science, as diverse chemical compositions can be achieved by a flexible combination of *A'*, *A* and *B* ions in the Dion–Jacobson phase.

Herein, we take attention to the cesium-containing Dion–Jacobson phase. As cesium has a very large ionic radius (1.74 Å for a coordination number of 8), we expected cesium-containing Dion–Jacobson phase to induce larger bottlenecks for oxide-ion diffusion. In the present study, we report oxide-ion conduction in the Dion–Jacobson phase $CsLa_2Ti_2NbO_{10}$ for the first time. Although $CsLa_2Ti_2NbO_{10}$ has been synthesized and the crystal structure was investigated by Young-Sik Hong in 2000, [32] their reported crystallographic data were incomplete and even have some mistakes. The *R* factors were high ($R_{wp} = 11.7\%$ and $R_p = 9.12\%$) and the isotropic atomic displacement parameter of oxygen (O4) was "– 0.1(3)". They just used the powder X-ray diffraction data at room temperature to do the refinement, which was difficult to confirm the states of light elements such as oxygen. Herein, we used the Rietveld analyses of neutron diffraction data to investigate the position and spatial distribution of the light element of oxygen. Furthermore, they did not consider the occupancy of each site of elements, and there was also no discussion on the anisotropic atomic displacement parameters. In our work, not only the room temperature crystal structure, but also the high-temperature structures were studied using high-temperature synchrotron X-ray powder diffraction and high-temperature neutron diffraction measurements. The occupancies and anisotropic atomic displacement parameters were also refined. At last, in order to improve the oxide-ion conductivity of $CsLa_2Ti_2NbO_{10}$, a lot of studies have been investigated.

3.2 Experimental

3.2.1 Synthesis and Characterization

$CsLa_2Ti_2NbO_{10}$, $CsLn_2Ti_2NbO_{10}$ (Ln = Pr, Nd, Sm, Gd, Ho, and Yb) and $CsLa_{2-x}M_xTi_{2+y}Nb_{1-y}O_{10-x/2-y/2}$ (M = Ca, Sr and Ba; $x = 0, 0.1$; $y = 0.1, 0$, and –0.1) were synthesized by the typical solid-state-reaction method. Mixtures of stoichiometric amounts of high-purity (>99.9%) Cs_2CO_3, La_2O_3, TiO_2, and Nb_2O_5 were ground using an agate mortar and a pestle for 2 h as ethanol slurries and as dried powders. The obtained mixtures were uniaxially pressed into pellets at about 100 MPa and then sintered in air at 1150 °C for 12 h with the heating and cooling rates of 5 °C min^{-1}. The processes of grind, pressing, and sintering were repeated two times. An excess amount of Cs_2CO_3 (30 mol%) was added to compensate for the mass loss due to the volatilization of the cesium (Cs) component. Parts of the sintered pellets of $CsLa_2Ti_2NbO_{10}$ were crushed and ground into powders to carry out X-ray powder diffraction (XRD, Rigaku RINT2550, Cu $K\alpha$ radiation), synchrotron X-ray powder diffraction (SXPRD), inductively coupled plasma optical emission spectroscopy ((ICP-OES, Hitachi PS3520UVDDII)), thermogravimetric (TG) analysis, and ultraviolet—visible (UV–vis) diffuse reflectance measurements. TG analyses were carried out in flowing dry and natural air between RT and 1073 K by the Bruker-AXS 2020SA instrument, where the heating and cooling rates were 5 K min^{-1}. The measurements were repeated three times to negate the influence of absorbed species such as water and also to confirm the reproducibility. The decomposition of the sample did not occur after the TG measurements, which were examined by XRD measurement. UV–vis diffuse reflectance spectrum was collected at room temperature and converted to absorbance over the spectral range of 250–850 nm (JASCO V-650 spectrometer).

3.2.2 Electrical Conductivity Measurements

The total electrical conductivity of $CsLa_2Ti_2NbO_{10}$, $CsLn_2Ti_2NbO_{10}$ (Ln = Pr, Nd, and Sm) and $CsLa_{2-x}M_xTi_{2+y}Nb_{1-y}O_{10-x/2-y/2}$ (M = Ca, Sr and Ba; $x = 0, 0.1$; $y = 0.1$, 0, and –0.1) were measured by a DC 4-probe method using sintered pellets (about 4.0 mm $\phi \times 10$ mm, relative density of 85%) with Pt electrodes in the temperature range between 400 and 1000 °C. The activation energy E_a was estimated from the Arrhenius equation

$$\sigma = \frac{A_0}{T}\exp\left(-\frac{E_a}{kT}\right)$$

where A_0, k and T are the pre-exponential factor, Boltzmann constant and absolute temperature, respectively. Temperature dependencies of conductivities of $CsLa_2Ti_2NbO_{10}$ were measured also in dry and wet air to investigate the proton conduction. Oxygen partial pressure $P(O_2)$ dependence of total electrical conductivities was carried out at 600, 800 and 1000 °C in the $P(O_2)$ region between 1×10^{-25} and 0 atm using N_2/H_2, N_2 and O_2 gas mixtures and the $P(O_2)$ was monitored by an oxygen sensor. Oxygen concentration-cell measurements were performed to determine the oxide-ion transport number (t_{ion}), using a sintered pellet (about 20 mm $\phi \times 3$ mm, relative density of 85%) attached to an alumina tube with a glass seal. One side of the pellet was exposed to a flowing dry air and the other side to a mixture of dry air and N_2. Electromotive forces of the concentration cell were measured by a Keithley model 617 electrometer. The following Nernst equation was used to estimate the t_{ion}, where F is the Faraday constant, R is the gas constant and T is the absolute temperature.

$$E = t_{ion}\frac{RT}{4F}ln\left(\frac{P(O_2)}{P^{\circ}(O_2)}\right)$$

3.2.3 *Neutron and Synchrotron X-ray Diffraction Measurements*

High-resolution SXRPD data of $CsLa_2Ti_2NbO_{10}$ were measured between 297 and 1073 K using six one-dimensional solid-state detectors at the BL02B2 beamline of SPring-8, Japan. Sieved powder samples were loaded into quartz glass capillaries ($\phi = 0.1$ mm) and sealed by clays. The wavelength was determined to be 0.6988697(5) Å using silicon standard powders (NIST SRM 640c). High-temperature neutron diffraction data were taken using the super-high-resolution diffractometer SuperHRPD installed at the Materials and Life Science Experimental Facility of the J-PARC center, Japan [33]. The diffraction data were analyzed by the Rietveld method using the Z-code program [34]. The absorption correction was performed using the method given by Rouse and Cooper. The refined crystal structure and bond-valence-based energy landscape were depicted by VESTA [35].

3.2.4 *Density Functional Theory Calculation*

Density functional theory (DFT) calculations of $CsLa_2Ti_2NbO_{10}$ were carried out to investigate the validity of refined crystal structure, band structure and the occupancy of Ti and Nb elements. Crystallographic parameters of $CsLa_2Ti_2NbO_{10}$ obtained from the Rietveld refinement of neutron-diffraction data were used as the initial model in the structural optimization. The generalized gradient approximation (GGA) electronic calculations were carried out with the Vienna Ab initio Simulation Package (VASP), in order to study the optimized structure of the 2 × 2 × 1 tetragonal supercells $CsLa_2Ti_2NbO_{10}$. Projector augmented-wave (PAW) potentials were used for Cs, La, Ti, Nb, and O atoms and a plane-wave basis set with a cutoff of 500 eV was used. The Perdew – Burke – Ernzerhof (PBE) GGA was used for the exchange and correlation functional. Unit-cell parameters and atomic coordinates were optimized with the convergence condition of 0.02 eV $Å^{-1}$. The self-consistent field tolerance of 1.0×10^{-7} eV/atom was used. The band structure obtained by the DFT calculations suggested that $CsLa_2Ti_2NbO_{10}$ has a direct bandgap. The 48 kinds of atomic configurations of $CsLa_2Ti_2NbO_{10}$ are shown later.

3.2.5 *Investigation of Oxide-Ion Diffusion Pathways*

Oxide-ion diffusion in the crystal structure is an important issue for an oxide-ion conductor. Bond valence-based energy landscapes (BVELs) for a test oxide ion of $CsLa_2Ti_2NbO_{10}$ were examined to investigate oxide-ion diffusion pathways and energy barriers for oxide-ion migration (E_b) using the neutron diffraction data with the computer program softBV [36, 37]. The bond valence parameters used in the calculation were given by Adams in the software [38]. The spatial resolution was set to 0.1 Å. The BVELs were drawn with VESTA [35].

3.2.6 *Improved Oxide-Ion Conductivity by Doping*

As discussed in Chap. 2, the oxide-ion conductivity can be improved by doping with increasing of the concentration of oxygen vacancies and or interstitial oxygen as well as to lower the activation energy by exchanging smaller cations with the larger cation. In the present work, $CsLa_{2-x}M_xTi_{2+y}Nb_{1-y}O_{10-x/2-y/2}$ (M = Ca, Sr and Ba; $x = 0, 0.1$; $y = 0.1, 0$, and -0.1) compounds were synthesized and the electrical properties were studied.

3.3 Results and Discussion

XRD patterns of $CsLn_2Ti_2NbO_{10}$ (Ln = La, Pr, Nd, Sm, Gd, Ho, and Yb) were shown in Fig. 3.1. Compared the XRD results with the reported literatures, the $CsLa_2Ti_2NbO_{10}$, $CsPr_2Ti_2NbO_{10}$, $CsNd_2Ti_2NbO_{10}$, and $CsSm_2Ti_2NbO_{10}$ were

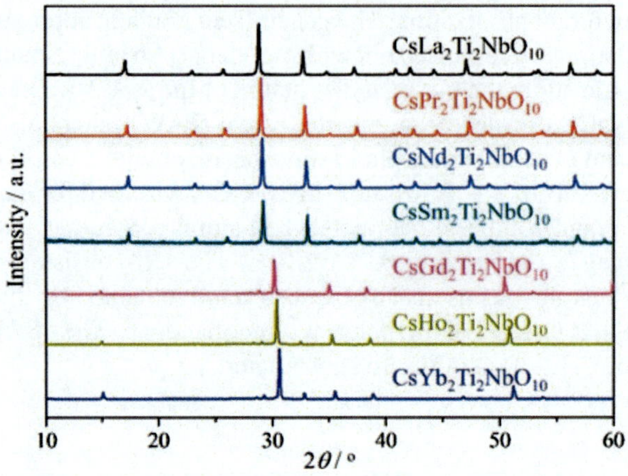

Fig. 3.1 X-ray powder diffraction patterns of $CsLn_2Ti_2NbO_{10}$ at room temperature

synthesized successfully. While the others just showed some impurities or just starting materials. Rietveld refinements of the synchrotron X-ray and neutron diffraction data of $CsLa_2Ti_2NbO_{10}$ were successfully performed using a single tetragonal phase with a space group of *P*4/*mmm* from RT to 800 °C without phase transition (Figs. 3.2, 3.3, 3.4, 3.5, 3.6, 3.7, 3.8, 3.9, 3.10, 3.11 and 3.12). The refined lattice parameters using neutron diffraction data were $a = b = 3.849759(2)$ Å and $c = 15.397263(17)$ Å at 24 °C, which is consistent with those from the literature [32].

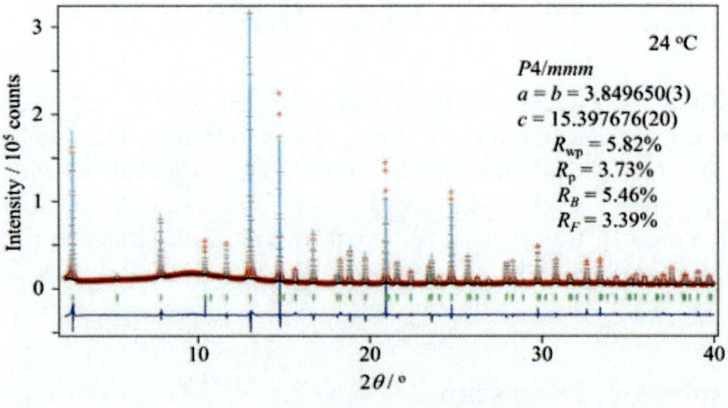

Fig. 3.2 Rietveld patterns of $CsLa_2Ti_2NbO_{10}$ taken at 297 K of synchrotron X-ray powder diffraction data measured at the BL02B2 station of SPring-8. The observed plots are shown by red marks, the calculated intensities are shown by light blue and the difference plots are shown by blue solid lines. Green tick marks stand for calculated Bragg peak positions

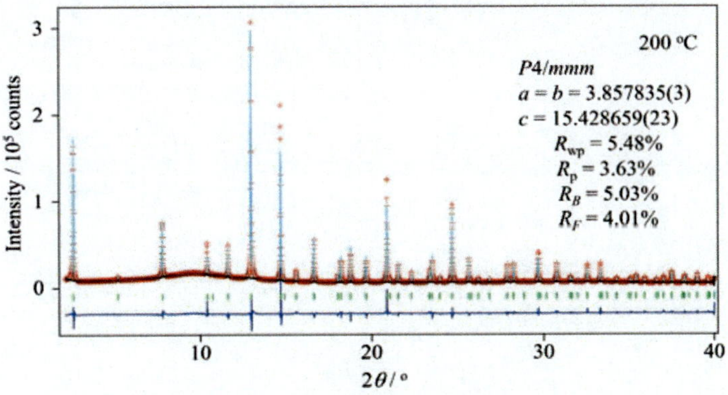

Fig. 3.3 Rietveld patterns of $CsLa_2Ti_2NbO_{10}$ taken at 473 K of synchrotron X-ray powder diffraction data measured at the BL02B2 station of SPring-8. The observed plots are shown by red marks, the calculated intensities are shown by light blue, and the difference plots are shown by blue solid lines. Green tick marks stand for calculated Bragg peak positions

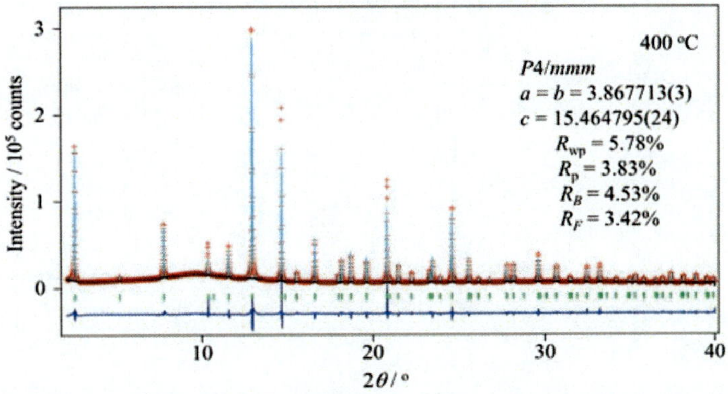

Fig. 3.4 Rietveld patterns of $CsLa_2Ti_2NbO_{10}$ taken at 673 K of synchrotron X-ray powder diffraction data measured at the BL02B2 station of SPring-8. The observed plots are shown by red marks, the calculated intensities are shown by light blue, and the difference plots are shown by blue solid lines. Green tick marks stand for calculated Bragg peak positions

Other refined crystal parameters at room temperature using neutron diffraction data were shown in Table 3.1. The refined crystal parameters of $CsLa_2Ti_2NbO_{10}$ obtained from synchrotron X-ray diffraction data agreed well with those from the neutron diffraction experiment (Fig. 3.13) and from the DFT-based structural optimization. The chemical composition of the synthesized product was determined to be Cs: La: Ti; Nb = 0.98: 2.03: 2.08: 1.01 [0.983(2): 2.034(12): 2.087(13): 1.008(5)] by ICP-OES, agreed well the nominal composition of Cs: La: Ti; Nb = 1: 2: 2: 1.

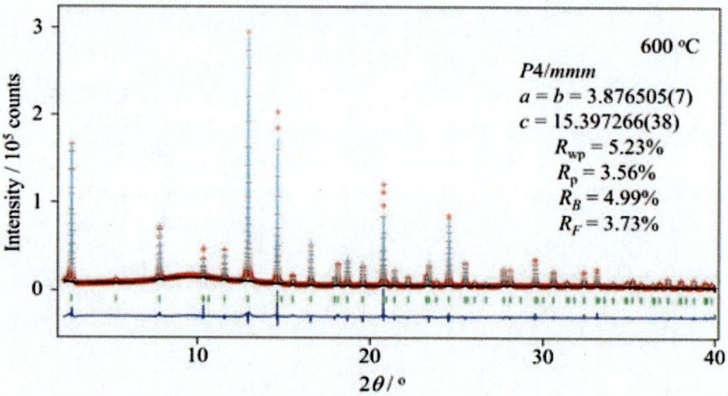

Fig. 3.5 Rietveld patterns of $CsLa_2Ti_2NbO_{10}$ taken at 873 K of synchrotron X-ray powder diffraction data measured at the BL02B2 station of SPring-8. The observed plots are shown by red marks, the calculated intensities are shown by light blue, and the difference plots are shown by blue solid lines. Green tick marks stand for calculated Bragg peak positions

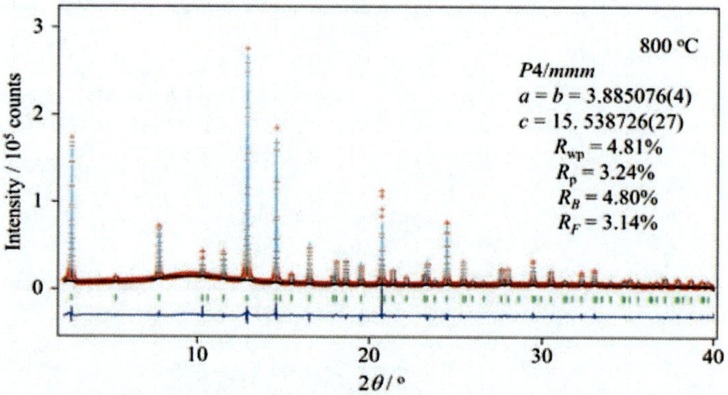

Fig. 3.6 Rietveld patterns of $CsLa_2Ti_2NbO_{10}$ taken at 1073 K of synchrotron X-ray powder diffraction data measured at the BL02B2 station of SPring-8. The observed plots are shown by red marks, the calculated intensities are shown by light blue, and the difference plots are shown by blue solid lines. Green tick marks stand for calculated Bragg peak positions

The bond valence sums (BVSs) of Cs, La, Ti/Nb, Ti, O1, O2, O3, and O4 sites in $CsLa_2Ti_2NbO_{10-\delta}$ at room temperature were calculated to be 1.05, 2.97, 4.50, 4.16, 2.02, 1.89, 2.16, and 1.96, respectively [39, 40]. The BVSs values agreed well with the formal charges, indicating the validity of the refined crystal structure of $CsLa_2Ti_2NbO_{10-\delta}$. The unit-cell parameters *a, b,* and *c,* increase linearly with an increase of temperature (Fig. 3.13) due to the thermal expansion. The average thermal expansion coefficients (TECs) of *a* (=*b*) and *c*-axis lengths between 40 and 800 °C were estimated using the refined neutron-diffraction data. The TECs

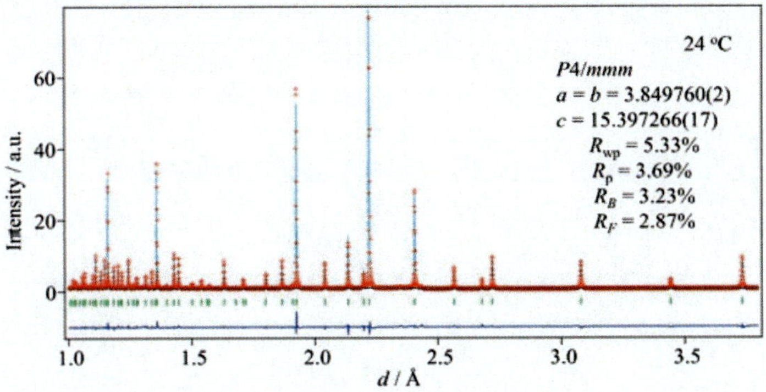

Fig. 3.7 Rietveld patterns of $CsLa_2Ti_2NbO_{10}$ taken at 297 K of neutron diffraction data taken by the backscattering bank of the Super-HRPD diffractometer. The observed plots are shown by red marks, the calculated intensities are shown by light blue, and the difference plots are shown by blue solid lines. Green tick marks stand for calculated Bragg peak positions

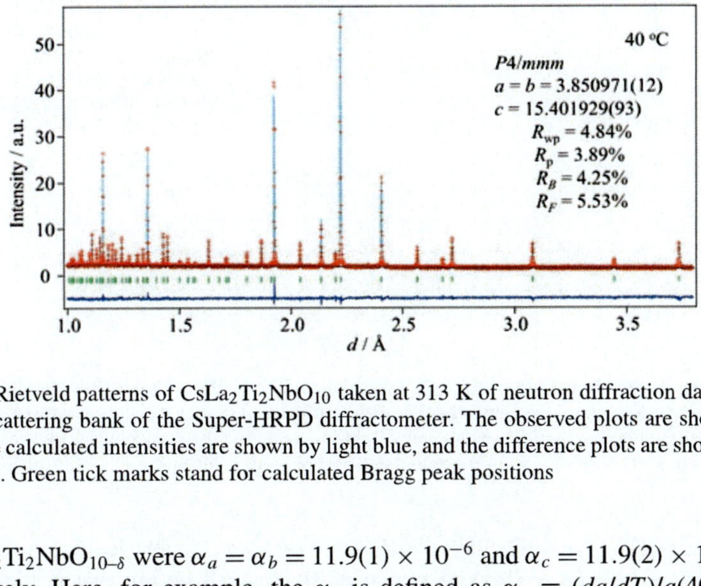

Fig. 3.8 Rietveld patterns of $CsLa_2Ti_2NbO_{10}$ taken at 313 K of neutron diffraction data taken by the backscattering bank of the Super-HRPD diffractometer. The observed plots are shown by red marks, the calculated intensities are shown by light blue, and the difference plots are shown by blue solid lines. Green tick marks stand for calculated Bragg peak positions

of $CsLa_2Ti_2NbO_{10-\delta}$ were $\alpha_a = \alpha_b = 11.9(1) \times 10^{-6}$ and $\alpha_c = 11.9(2) \times 10^{-6}$ K^{-1}, respectively. Here, for example, the α_a is defined as $\alpha_a \equiv (da/dT)/a(40)$, where $a(40)$ is the unit-cell parameter a at 40 °C and da/dT is the slope of the linear line in Fig. 3.13 (neutron-diffraction data). The average linear thermal expansion coefficient of $CsLa_2Ti_2NbO_{10-\delta}$ was calculated to be $11.9(2) \times 10^{-6}$ K^{-1}, which is close to that of 8 mol % $Y_2O_3 - ZrO_2$ (10.5×10^{-6} K^{-1}).

Figure 3.14 shows the refined crystal structure of $CsLa_2Ti_2NbO_{10}$, which was obtained by the Rietveld analysis of the neutron-diffraction data measured at 24 °C. The $CsLa_2Ti_2NbO_{10}$ phase consists of the inner perovskite $LaTiO_3$ layer, outer

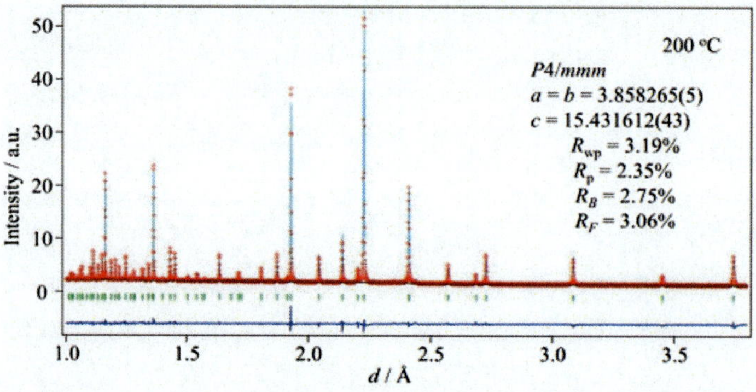

Fig. 3.9 Rietveld patterns of $CsLa_2Ti_2NbO_{10}$ taken at 473 K of neutron diffraction data taken by the backscattering bank of the Super-HRPD diffractometer. The observed plots are shown by red marks, the calculated intensities are shown by light blue, and the difference plots are shown by blue solid lines. Green tick marks stand for calculated Bragg peak positions

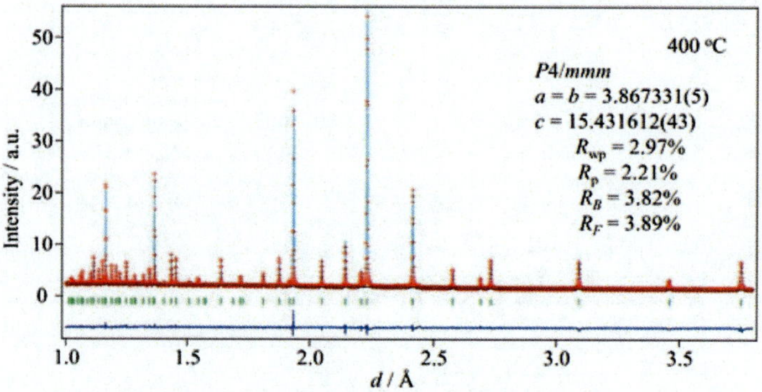

Fig. 3.10 Rietveld patterns of $CsLa_2Ti_2NbO_{10}$ taken at 673 K of neutron diffraction data taken by the backscattering bank of the Super-HRPD diffractometer. The observed plots are shown by red marks, the calculated intensities are shown by light blue, and the difference plots are shown by blue solid lines. Green tick marks stand for calculated Bragg peak positions

perovskite $La(Ti_{0.5}Nb_{0.5})O_3$ layer and CsO rock-salt layer, leading to a Dion–Jacobson-type layered perovskite structure. It is worth noting that Nb cation just occupies the *B*-site of the outer perovskite layer. In the previous report, they have performed just 3 models to discuss the occupation of Ti and Nb cations. Herein, we have verified 48 models using DFT calculations to confirm the occupation of Ti and Nb sites, combining with the refinement results of the synchrotron X-ray and neutron diffraction data. The bond valence sums (BVSs) of the Ti-site and Ti/Nb site were estimated to be 4.16 and 4.50, which also indicates the validity of the occupation.

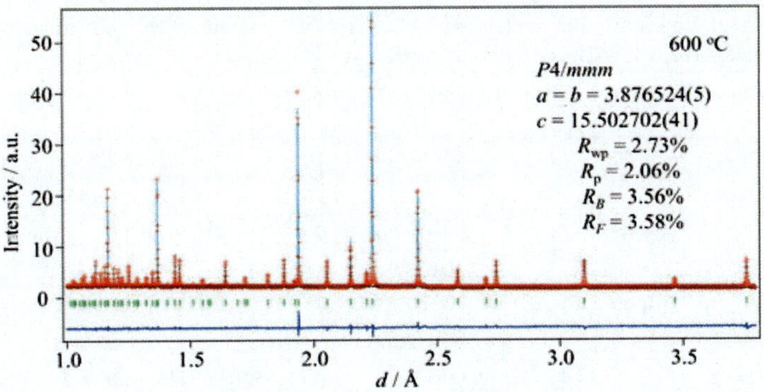

Fig. 3.11 Rietveld patterns of $CsLa_2Ti_2NbO_{10}$ taken at 873 K of neutron diffraction data taken by the backscattering bank of the Super-HRPD diffractometer. The observed plots are shown by red marks, the calculated intensities are shown by light blue, and the difference plots are shown by blue solid lines. Green tick marks stand for calculated Bragg peak positions

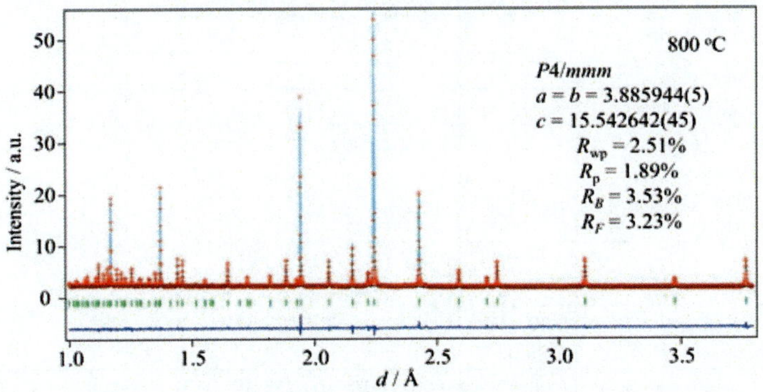

Fig. 3.12 Rietveld patterns of $CsLa_2Ti_2NbO_{10}$ taken at 1073 K of neutron diffraction data taken by the backscattering bank of the Super-HRPD diffractometer. The observed plots are shown by red marks, the calculated intensities are shown by light blue, and the difference plots are shown by blue solid lines. Green tick marks stand for calculated Bragg peak positions

Optical bandgap E_g of $CsLa_2Ti_2NbO_{10-\delta}$ was estimated to be 3.85 eV from the Tauc plot for the direct bandgap from the UV – vis reflectance spectrum over the spectral range of 200–850 nm (Fig. 3.15), which was confirmed by the DFT calculations Fig. 3.16. The wide bandgap shows semi-conduction in $CsLa_2Ti_2NbO_{10-\delta}$. The bond valence energy landscapes of Cs^{1+} and O^{2-} ions were calculated in $CsLa_2Ti_2NbO_{10-\delta}$. The energy barriers for the Cs^{1+} and O^{2-} migration were 5.4 eV and 0.86 eV, respectively, which suggests the oxide-ion conduction is dominant in $CsLa_2Ti_2NbO_{10-\delta}$.

Table 3.1 Crystallographic parameters of $CsLa_2Ti_2NbO_{10}$ which were refined using neutron-diffraction data taken with SuperHRPD diffraction at 24 °C

Site s	Atom X oxidation number		Wyckoff position	g(X; s) [a]	X	y	z	BVS[b]
Cs	Cs + 1		1*d*	1	1/2	1/2	1/2	0.96
La	La + 3		2 *h*	1	1/2	1/2	0.14042(2)	3.10
Ti/Nb1	Ti	+ 4.02	1*a*	0.9847(5)	0	0	0	4.08
	Nb			0.0153(5)				
Ti/Nb2	Ti	+ 4.49	2 *g*	0.5077(2)	0	0	0.28115(7)	4.46
	Nb			0.4923(2)				
O1	O − 2		4*n*	0.5	0.11436(9)	1/2	0	2.05
O2	O − 2		2 *g*	1	0	0	0.12410(2)	1.94
O3	O − 2		4*i*	1	0	1/2	0.25004(2)	2.14
O4	O − 2		2 *g*	1	0	0	0.39492(2)	1.83
Site *s*	U_{iso} [c] or U_{eq} [d] (Å^2)		U_{11} (Å^2) [e]	U_{33} (Å^2) [e]	U_{12} (Å^2) [e]	U_{13} (Å^2) [e]	U_{13} (Å^2) [e]	U_{23} (Å^2) [e]
Cs	U_{eq} = 0.01403(15)		0.0072(2)	0.0072(2)	0.0277(5)	0	0	0
La	U_{iso} = 0.00240(6)		–	–	–	–	–	–
Ti/Nb1	U_{iso} = 0.0023(2)		–	–	–	–	–	–
Ti/Nb2	= U_{iso}(Ti; Ti/Nb2) = U_{iso}(Nb; Ti/Nb2)		–	–	–	–	–	–
O1	U_{eq} = 0.01010(16)		0.0010(4)	0.0011(3)	0.0283(4)	0	0	0
O2	U_{eq} = 0.01248(11)		0.0073(2)	0.0073(2)	0.0228(4)	0	0	0
O3	U_{eq} = 0.00715(6)		0.00297(10)	0.00297(10)	0.0155(3)	0	0	0
O4	U_{eq} = 0.00891(9)		0.00877(14)	0.00877(14)	0.0092(3)	0	0	0

Crystal system: tetragonal; Space group: *P4/mmm*; Lattice parameters: $a = b = 3.849762(2)$ Å and $c = 15.39727(2)$ Å

Reliability factors: $R_{wp} = 5.35\%$, $R_p = 3.69\%$, $R_B = 3.22\%$, and $R_F = 3.07\%$

[a] *g(X; s)*: Occupancy factor of *X* atom at the *s* site

[b] Bond valence sums (BVSs) were calculated using bond valence parameters from references

[c] $U_{iso}(X; s)$: Isotropic atomic displacement parameter of *X* atom at the *s* site

[d] U_{eq}: Equivalent isotropic atomic displacement parameter

[e] U_{ij}: Anisotropic atomic displacement parameter

[f] The following linear constraints were used in the refinement: U_{iso}(Ti; Ti/Nb1) = U_{iso}(Nb; Ti/Nb1) = U_{iso}(Ti; Ti/Nb2) = U_{iso}(Nb; Ti/Nb2)

[g] The occupancy factors of Cs, La, O1 × 2, O2, O3, and O4 were fixed to be 1 because the refined value was 1 within 3 times of the estimated deviation or higher than 1

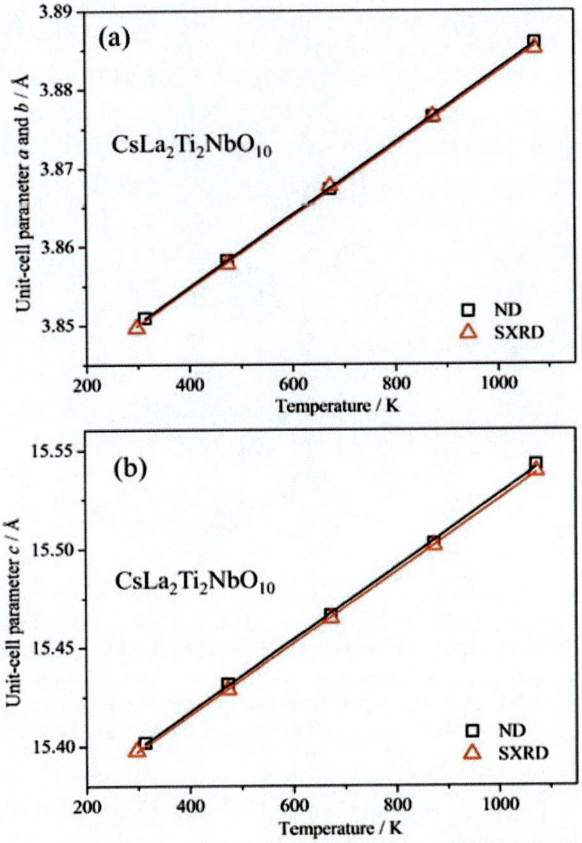

Fig. 3.13 Temperature dependence of unit-cell parameters (**a**) $a = b$ and (**b**) c, refined from the time-of-flight neutron diffraction data taken by the backscattering bank of the Super-HRPD diffractometer and the synchrotron X-ray powder diffraction data measured at the BL02B2 station of SPring-8

The total electrical conductivities of $CsLa_2Ti_2NbO_{10}$, $CsPr_2Ti_2NbO_{10}$, $CsNd_2Ti_2NbO_{10}$, and $CsSm_2Ti_2NbO_{10}$ were measured from 573 to 1173 K using DC methods (Fig. 3.17). the $CsLa_2Ti_2NbO_{10}$ showed the highest total electrical conductivity with an activation energy E_a of 0.78(2) eV. In addition, we have also measured the conductivity of $CsLa_2Ti_2NbO_{10}$ and $CsPr_2Ti_2NbO_{10}$ at low-temperature ranges. (Fig. 3.18) The total electrical conductivities decrease with the increasing temperature above 400 K, there may exist a slight level of proton conduction.

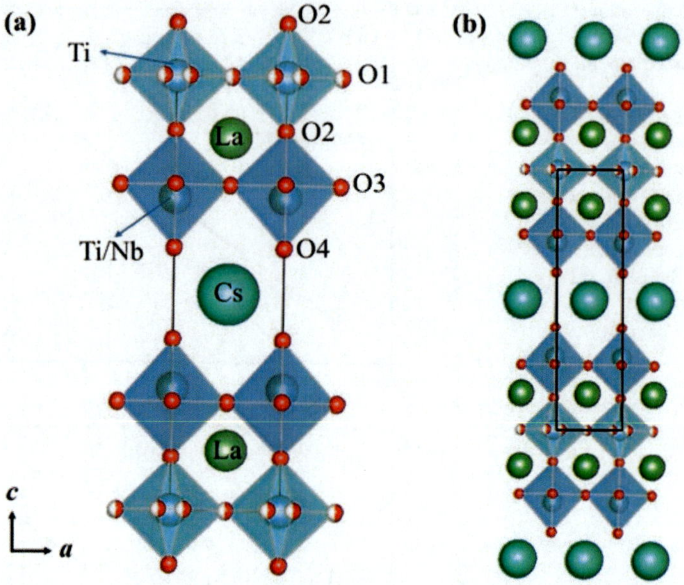

Fig. 3.14 Refined crystal structure of $CsLa_2Ti_2NbO_{10}$ at 24 °C, which were obtained by Rietveld analyses of in situ neutron-diffraction data. Thermal ellipsoids are drawn at the 50% probability level. Ranges of fractional coordinates in drawing the crystal structure: $0 \leq x \leq 1, 0 \leq y \leq 1, 0 \leq z \leq 1$ for (**a**) and $0 \leq x \leq 1, 0 \leq y \leq 1, -0.5 \leq z \leq 1.5$ for (**b**). Solid lines denote the unit cells

Oxygen concentration-cell measurements of $CsLa_2Ti_2NbO_{10}$ were performed to determine the oxide-ion transport number t_{ion}. The oxide-ion transport number (Fig. 3.19) indicates that the dominant carrier in the Dion–Jacobson phase $CsLa_2Ti_2NbO_{10-\delta}$ is an oxide ion. The total electrical conductivities are almost constant in the oxygen partial pressure $P(O_2)$ region around 1×10^{-20} and 0 atm (Fig. 3.20), supporting the oxide-ion conduction. Furthermore, no significant proton conduction was observed between 400 and 1000 °C, because the conductivities measured in dry air agreed well with those in wet air (Fig. 3.21). The total electrical conductivity increased with an increase in temperature. The activation energies E_a for the total electrical conductivity were calculated to be 0.78(2) eV from the Arrhenius plot (Fig. 3.17), which was comparatively low during the perovskite-related oxide-ion conductors. We attributed the low E_a to the large bottlenecks for oxide-ion migration as described later. The oxide-ion conductivities σ_{ion} of $CsLa_2Ti_2NbO_{10-\delta}$ were estimated using an equation $\sigma_{\text{ion}} = \sigma_{\text{tot}} \times t_{\text{ion}}$. The σ_{ion} increased with an increase in temperature and the activation energy E_a (oxide-ion) was calculated to

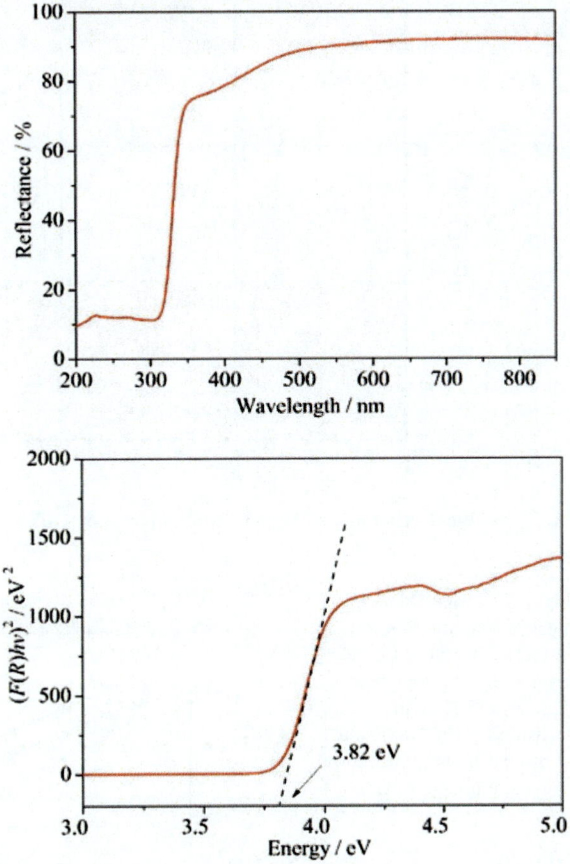

Fig. 3.15 **a** Diffuse reflectance spectrum and **b** Tauc plot for the direct bandgap of $CsLa_2Ti_2NbO_{10}$

be 0.878(2) eV (Fig. 3.22). The pellet after $P(O_2)$ measurements was crushed and ground into powder. The powder was used for the X-ray diffraction measurements in order to investigate the phase stability at high temperatures and different atmospheres. There was no change before and after the measurements in the X-ray powder diffraction patterns (Fig. 3.23), which demonstrated the high phase stability at high temperatures and different $P(O_2)$.

In order to investigate the oxide-ion migration pathway of $CsLa_2Ti_2NbO_{10}$, the BVE of an oxide ion was calculated using the refined crystal structure of $CsLa_2Ti_2NbO_{10}$ at 24 °C. Figures 3.24 and 3.25 show the yellow isosurfaces of BVE for an oxide ion at + 1.6 eV. In these BVE landscapes, the most stable position of an oxide ion was set to 0 eV. Four oxide-ion migration paths were found on the *ab* plane. The energy barriers through O1 – O1 path, O1 – O2 path, O2 – O3 path and O3 – O3 path were 0.85 eV, 1.44 eV, 1.53e V and 1.36e V, respectively. In addition to these paths, the energy barrier of oxide ion migration along the *c*-axis was 4.81 eV.

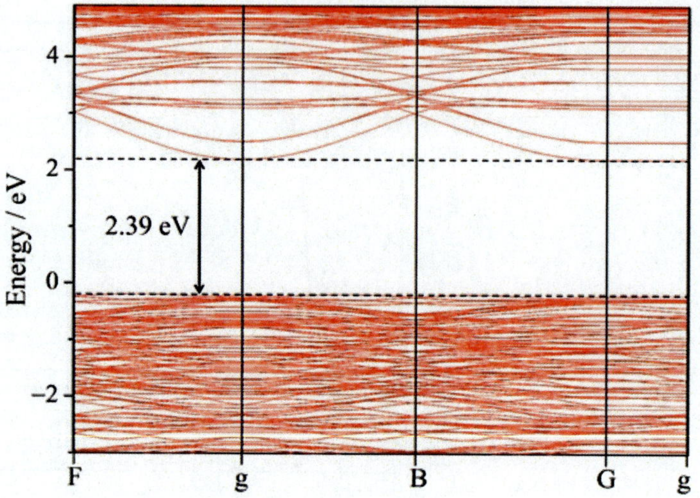

Fig. 3.16 Band structure of $CsLa_2Ti_2NbO_{10}$ based on DFT calculations

Fig. 3.17 Total DC electrical conductivities σ_{tot} of $CsLn_2Ti_2NbO_{10}$ (Ln = La, Pr, Nd, and Sm) from 673 to 1273 K in air, measured every 100 K and allowed 1 h of equilibration at each temperature step

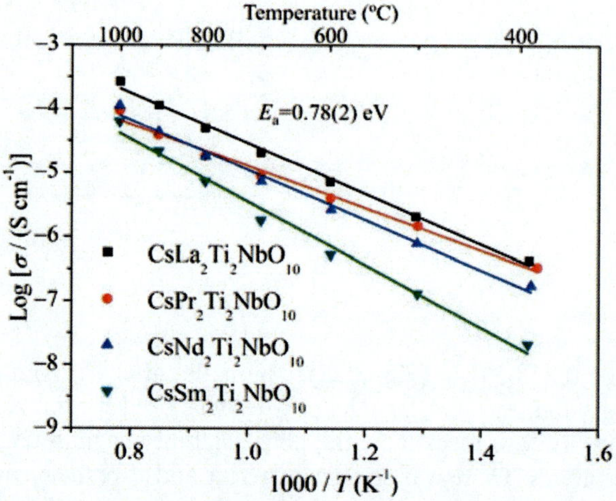

The energy barriers decreased with increasing temperature as shown in Fig. 3.26. Therefore, it is strongly suggested the two-dimensional oxide ions diffusion path in the $CsLa_2Ti_2NbO_{10}$.

Aliovalent cations substitution and doping using different cations are effective to improve the oxide-ion conductivities, which can increase either oxygen vacancies or interstitial oxygen and by decreasing activation energy for oxide-ion conduction. In the present work, the $CsLa_{2-x}M_xTi_{2+y}Nb_{1-y}O_{10-x/2-y/2}$ (M = Ca, Sr, and Ba; x = 0, 0.1; y = 0.1, 0, and −0.1) compounds were investigated (Fig. 3.27). The total electrical conductivity was given in Fig. 3.28, the $CsLa_2Ti_{2.1}Nb_{0.9}O_{9.95}$ showed high

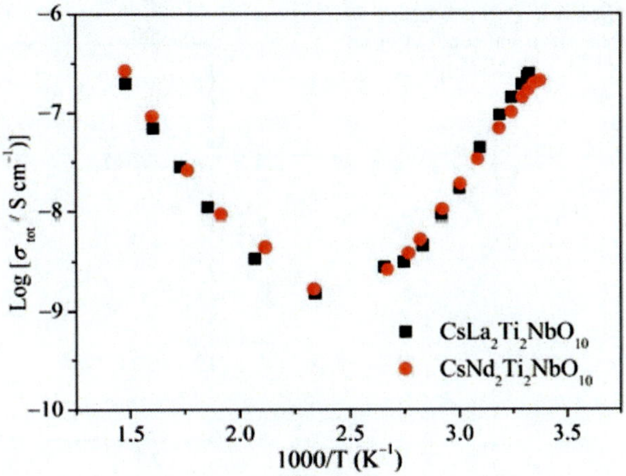

Fig. 3.18 Total DC electrical conductivities σ_{tot} of $CsLn_2Ti_2NbO_{10}$ (Ln = La and Nd) from 300 to 673 K in air

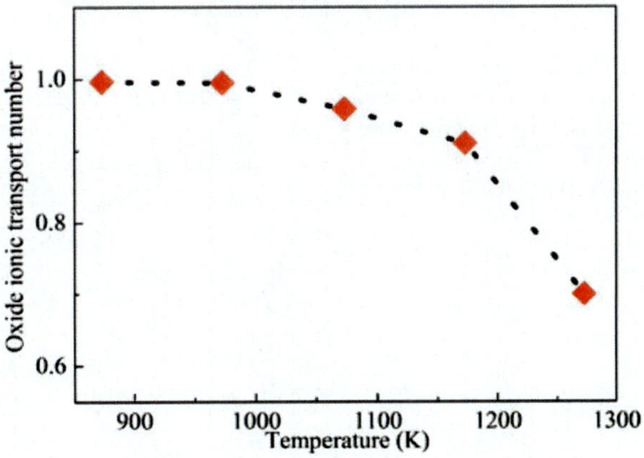

Fig. 3.19 Oxide-ion transport number determined from oxygen concentration-cell measurements of $CsLa_2Ti_2NbO_{10-\delta}$

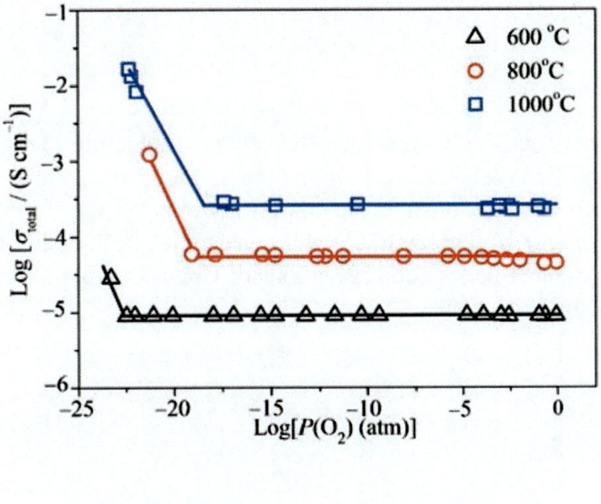

Fig. 3.20 Partial oxygen pressure $P(O_2)$ dependence of the total electrical conductivity of $CsLa_2Ti_2NbO_{10-\delta}$ at 600, 800 and 1000 °C

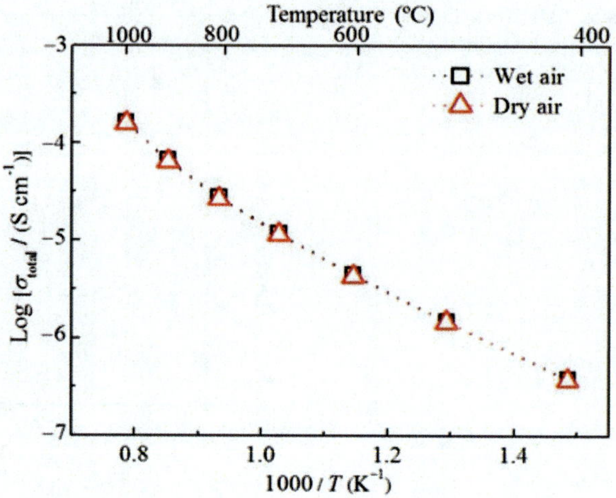

Fig. 3.21 Arrhenius plots of total electrical conductivities of $CsLa_2Ti_2NbO_{10-\delta}$ in dry and wet air

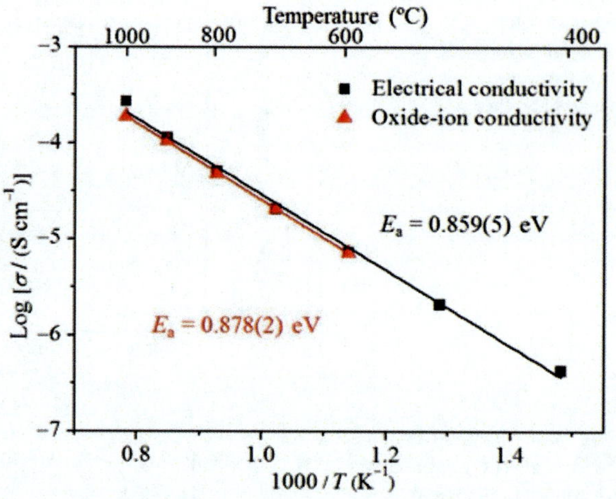

Fig. 3.22 Arrhenius plots of total electrical conductivities and oxide-ion conductivities of $CsLa_2Ti_2NbO_{10-\delta}$. Oxide-ion conductivities were calculated using the equation of $\sigma_{ion} = \sigma_{tot} \times t_{ion}$

conductivity than the mother phase, indicating the success of the doping method, while others showed almost the same or even lower conductivity than the mother phase. We consider the impurities that existed during the sintering and low level of carrier charges like oxygen vacancies or interstitial oxygen atom. At last, the 48 kinds of atomic configurations of $CsLa_2Ti_2NbO_{10}$ in DFT calculation are shown in Fig. 3.29.

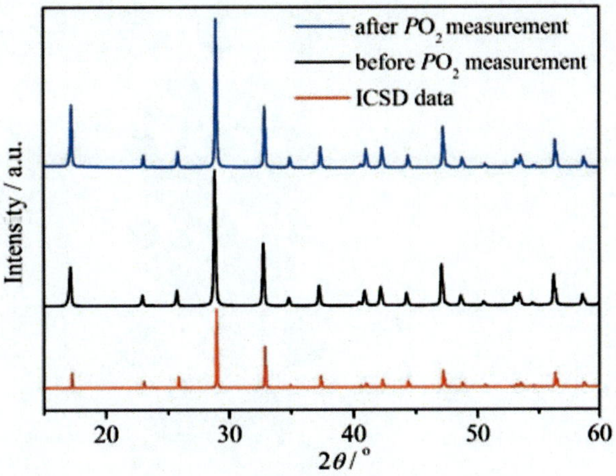

Fig. 3.23 X-ray powder diffraction patterns of CsNb Ti_2NbO_{10} before and after $P(O_2)$ measurements

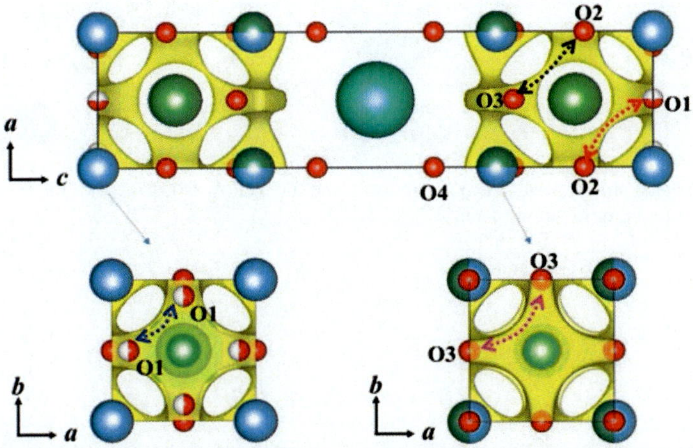

Fig. 3.24 Yellow isosurfaces of the bond-valence-based energy for an oxide ion at 1.6 eV in $CsLa_2Ti_2NbO_{10-\delta}$. Dotted lines with arrows denote the possible diffusion pathways of oxide ion. Thermal ellipsoids are drawn at the 50% probability level

The present work has demonstrated the oxide-ion conductivity in the Cs-containing Dion–Jacobson phase $CsLa_2Ti_2NbO_{10-\delta}$, with a comparable low activation energy. We attribute this to enlarged bottlenecks for oxide-ion migration through a large ionic radius of Cs^+. A larger radius of Cs makes the longer La–La and La–Ti distances. The longer La–La and La–Ti distances lead to larger bottlenecks for oxide ion migration.

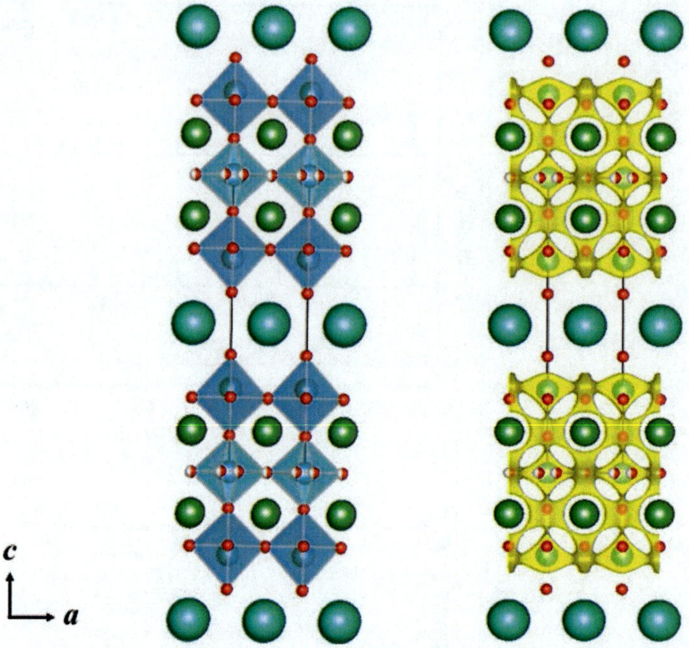

Fig. 3.25 Crystal structure of $CsLa_2Ti_2NbO_{10-\delta}$ and the Yellow isosurfaces of the bond-valence-based energy for an oxide ion at 1.6 eV

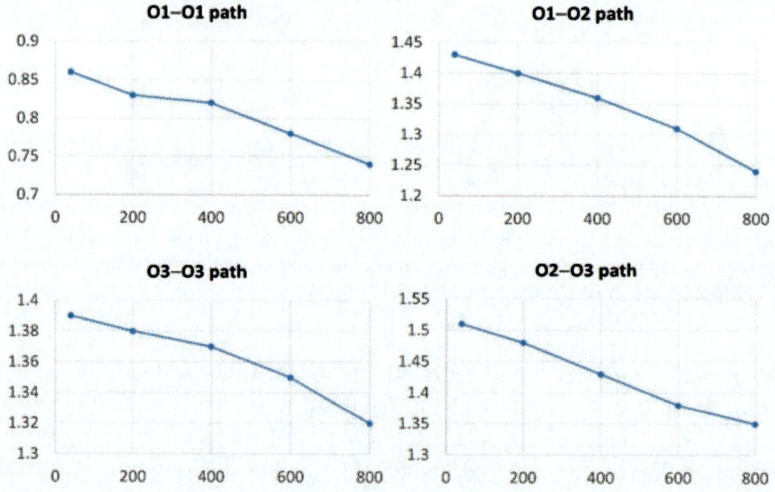

Fig. 3.26 Temperature dependence of energy barrier for oxide ion migration in $CsLa_2Ti_2NbO_{10-\delta}$

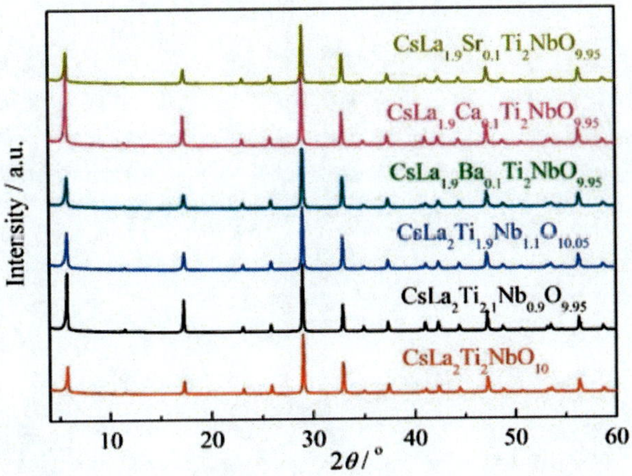

Fig. 3.27 X-ray powder diffraction patterns of $CsLa_{2-x}M_xTi_{2+y}Nb_{1-y}O_{10-x/2-y/2}$ (M = Ca, Sr, and Ba; $x = 0, 0.1$; $y = 0.1, 0$, and -0.1)

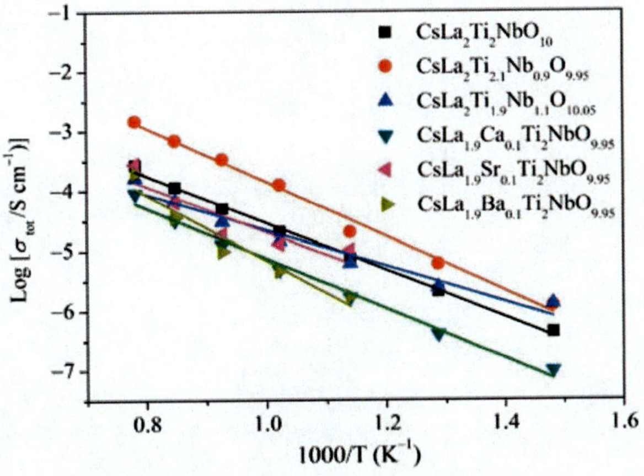

Fig. 3.28 Total DC electrical conductivities σ_{tot} of $CsLa_{2-x}M_xTi_{2+y}Nb_{1-y}O_{10-x/2-y/2}$ (M = Ca, Sr, and Ba; $x = 0, 0.1$; $y = 0.1, 0$, and -0.1)

3.4 Conclusion

In summary, we have discovered the Dion–Jacobson phase $CsLa_2Ti_2NbO_{10-\delta}$ as a new structure-type oxide-ion conductor. In addition, some of its related oxide and doping compounds were also synthesized and investigated. The bond-valence-based energy landscapes of a test oxide ion in $CsLa_2Ti_2NbO_{10-\delta}$ showed two-dimensional oxide-ion diffusion along the edges of an octahedron in the perovskite layers. We believe that better oxide-ion conductors can be designed by doping modifications in the basic material of $CsLa_2Ti_2NbO_{10-\delta}$. Compared with the $CsBi_2Ti_2NbO_{10-\delta}$

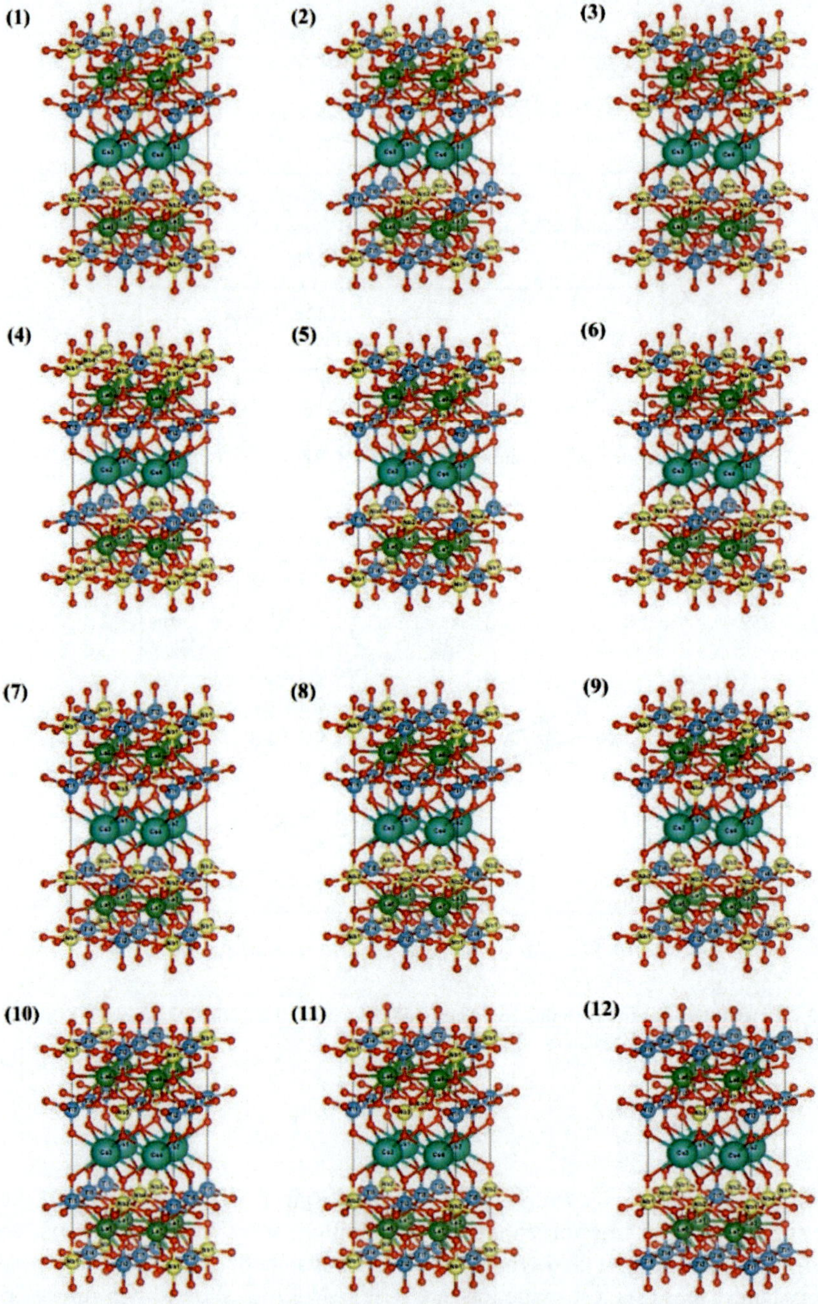

Fig. 3.29 48 kinds of structure models of $CsLa_2Ti_2NbO_{10}$ for DFT-based structural optimizations to investigate the occupancy of Ti and Nb cations in the *B*-site

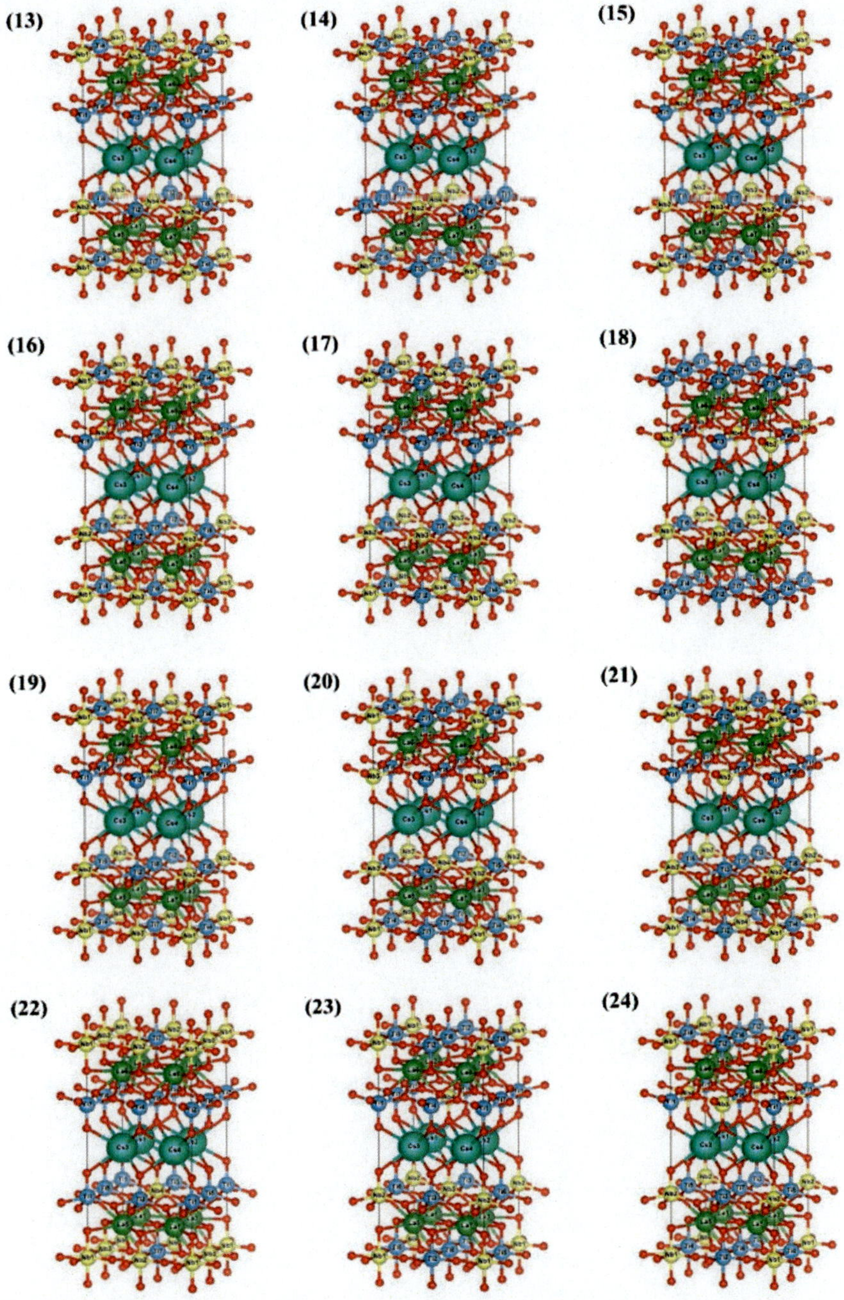

Fig. 3.29 (continued)

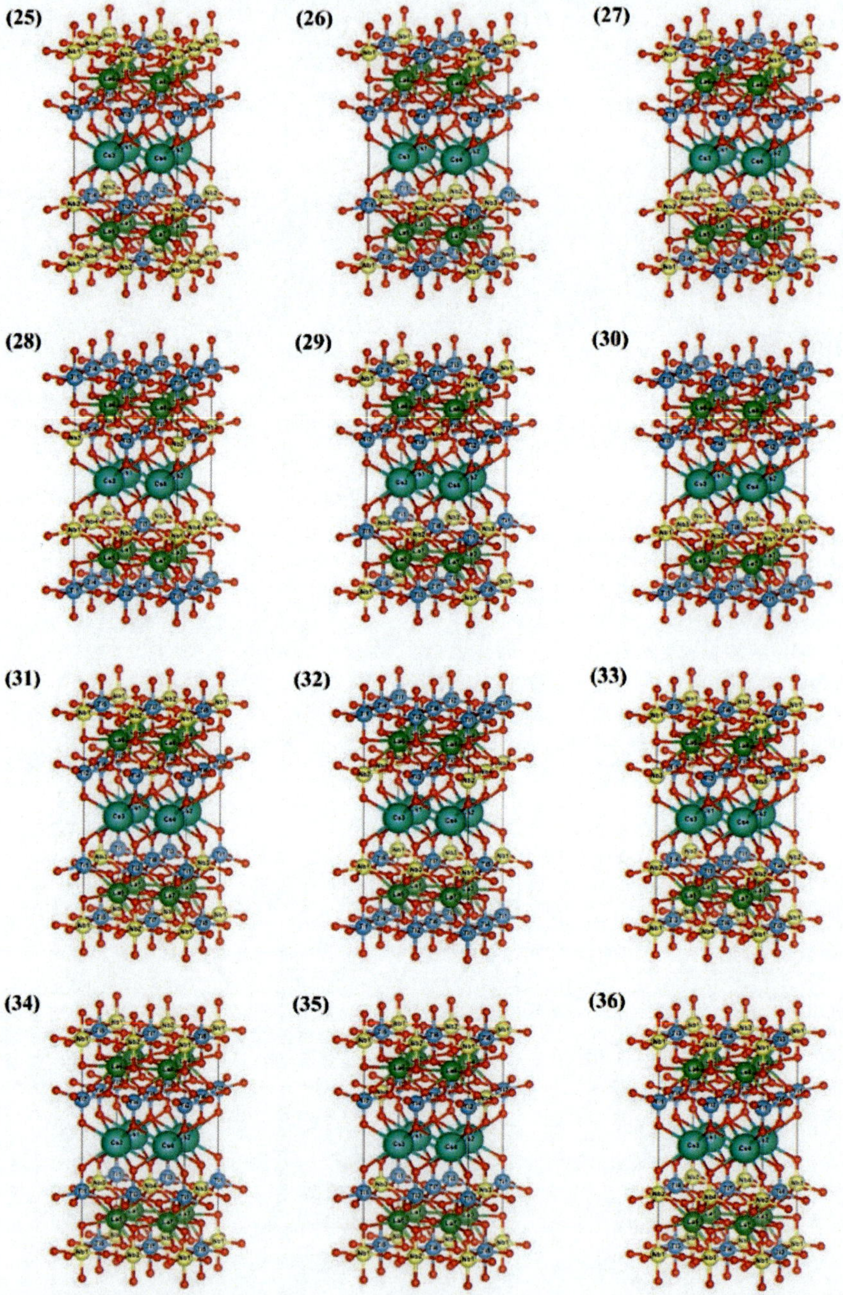

Fig. 3.29 (continued)

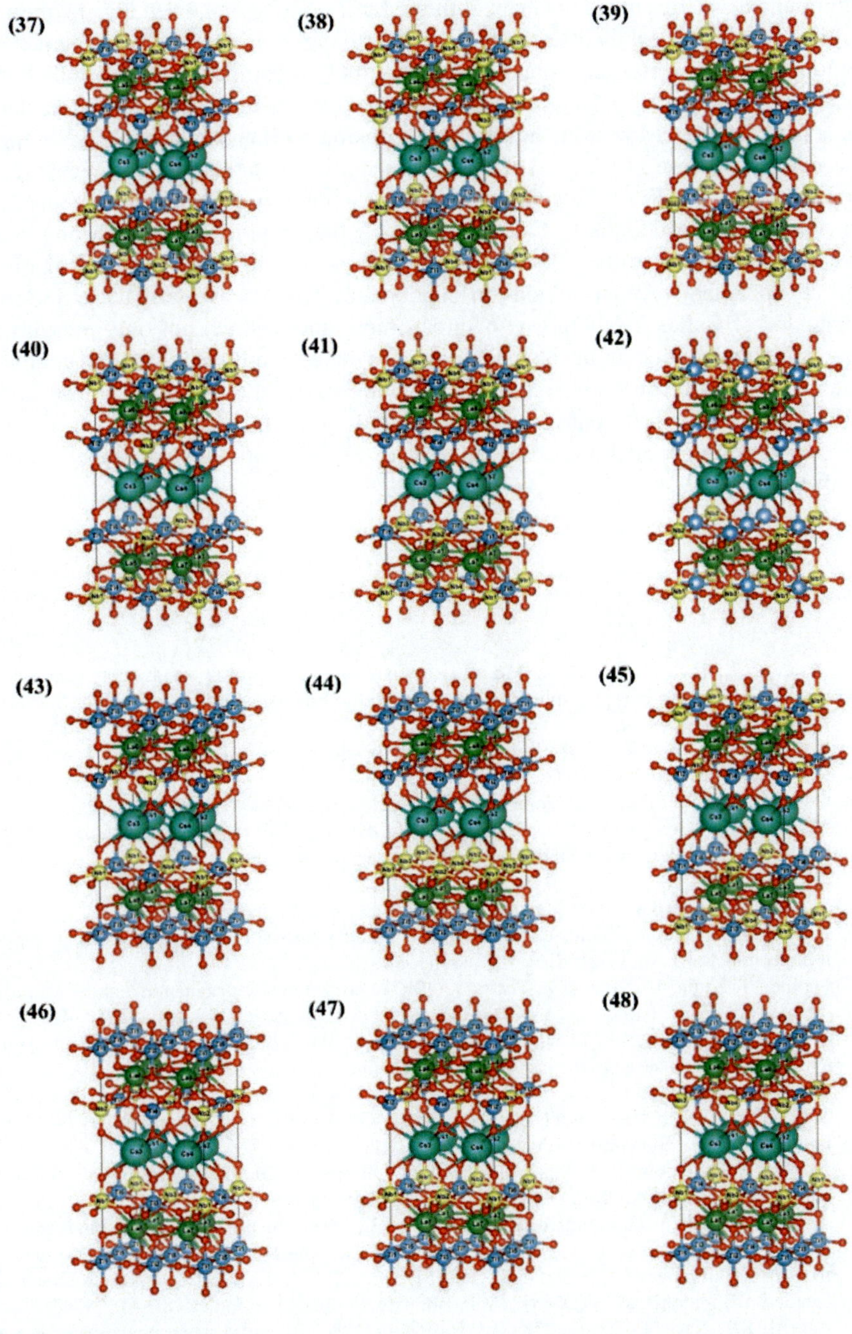

Fig. 3.29 (continued)

compound as we discussed in Chap. 2, the $CsLa_2Ti_2NbO_{10-\delta}$ showed lower conductivities, which is attributed to the low level of oxygen vacancies (carrier concentration). However, different from the $CsBi_2Ti_2NbO_{10-\delta}$ compound, there is no phase transition in the $CsLa_2Ti_2NbO_{10-\delta}$, which showed the practice application as an electrode in the SOFCs. Moreover, the thermal expansion coefficient of $CsLa_2Ti_2NbO_{10-\delta}$ was calculated to be $11.9(2) \times 10^{-6}$ K^{-1}, which is close to that of 8 mol % $Y_2O_3 - ZrO_2$ (10.5×10^{-6} K^{-1}). This study demonstrates not only an excellent candidate as a new oxide-ion conductor of $CsLa_2Ti_2NbO_{10-\delta}$, but also the Dion–Jacobson phase as a new structural family of oxide-ion conductors, leading to further investigation and advancement of novel oxide-ion conductors. The discovery of the oxide-ion conductor of $CsLa_2Ti_2NbO_{10-\delta}$ is of great importance, which not only presents a new oxide-ion conductor, but also demonstrates the possibility of oxide-ion conduction in the Dion–Jacobson phase. Since the wide range of the chemical compositions in the Dion – Jacobson phases with the formula of A' $[A_{n-1}B_nO_{3n+1}]$, we believe novel oxide-ion conductors can be discovered in the Dion–Jacobson type layered perovskites.

References

1. Malavasi L, Fisher CAJ, Islam MS (2010) Oxide-ion and proton conducting electrolyte materials for clean energy applications: structural and mechanistic features. Chem Soc Rev 39(11):4370–4387
2. Boivin JC, Mairesse G (1998) Recent material developments in fast oxide-ion conductors. Chem Mater 10(10):2870–2888
3. Goodenough JB (2000) Oxide-ion conductors by design. Nature 404(6780):821–823
4. Kharton VV, Yaremchenko AA, Naumovich EN, Marques FMB (1999) Research on the electrochemistry of oxygen ion conductors in the former Soviet Union. J Solid State Electrochem 3:303–326
5. Skinner SJ, Kilner JA (2003) Ion conductors. Rev Lit Arts Am 6(3):30–37
6. Ishihara T, Matsuda H, Takita Y (1994) Doped LaGaO3 perovskite type oxide as a new oxide ionic conductor. J Am Chem Soc 116(9):3801–3803
7. Ishihara T, Kilner JA, Honda M, Takita Y (1997) Oxygen surface exchange and diffusion in the new perovskite oxide-ion conductor LaGaO3. J Am Chem Soc 119(11):2747–2748
8. Malavasi L, Kim HJ, Billinge SJL, Proffen T, Tealdi C, Flor G (2007) Nature of the monoclinic to cubic phase transition in the fast oxygen ion conductor La2Mo2O9 (LAMOX). J Am Chem Soc 129(21):6903–6907
9. Manthiram A, Kim JH, Kim YN, Lee KT (2011) Crystal chemistry and properties of mixed ionic-electronic conductors. J Electroceramics 27(2):93–107
10. Lacorre P, Goutenoire Ë, Bohnke O, Retoux R, Laligant Y (2000) Designing fast oxide-ion conductors based on La2Mo2O9. Nature 404, no. April, p 856
11. Abraham F, Boivin JC, Mairesse G, Nowogrocki G (1990) The bimevox series: a new family of high performances oxide-ion conductors. Solid State Ionics 40–41, no. PART 2:934–937
12. Iharada T, Hammouche A, Fouletier J, Kleitz M, Boivin JC, Mairesse G (1991) Electrochemical characterization of BIMEVOX oxide-ion conductors. Solid State Ionics 48(3–4):257–265
13. Kendall KR, Navas C, Thomas JK, Zur Loye HC (1996) Recent developments in oxide-ion conductors: Aurivillius phases. Chem Mater 8(3):642–649
14. Thomas JK, Anderson ME, Krause WE, Zur Loye H (1993) Oxygen ion conductivity in a new class of layered bismuth oxide compounds. MRS Online Proc Libr Arch 293:295–300

15. Amow G, Skinner SJ (2006) Recent developments in Ruddlesden-Popper nickelate systems for solid oxide fuel cell cathodes. J Solid State Electrochem 10(8):538–546
16. Patrakeev MV, Leonidov IA, Kozhevnikov VL, Kharton VV (2004) Ion-electron transport in strontium ferrites: relationships with structural features and stability. Solid State Sci 6(9):907–913
17. Yamada A et al (2008) Ruddlesden-popper-type epitaxial film as oxygen electrode for solid-oxide fuel cells. Adv Mater 20(21):4124–4128
18. Sengodan S et al (2015) Layered oxygen-deficient double perovskite as an efficient and stable anode for direct hydrocarbon solid oxide fuel cells. Nat Mater 14(2):205–209
19. Ali R, Yashima M, Izumi F (2007) Diffusion path of oxide ions in an oxide-ion conductor la 0.64(Ti0.92Nb0.08)O2.99 with a double perovskite-type structure. Chem Mater 19(13):3260–3264
20. Goodenough JB, Ruiz-Diaz JE, Zhen YS (1990) Oxide-ion conduction in Ba2In2O5 and Ba3In2MO8 (M=Ce, Hf, or Zr). Solid State Ionics 44(1–2):21–31
21. Auckett JE et al (2013) Combined experimental and computational study of oxide ion conduction dynamics in Sr2Fe2O5 brownmillerite. Chem Mater 25(15):3080–3087
22. Fop S, McCombie KS, Wildman EJ, Skakle JMS, Mclaughlin AC (2019) Hexagonal perovskite derivatives: a new direction in the design of oxide ion conducting materials. Chem Commun 55(15):2127–2137
23. Fop S et al (2016) Oxide ion conductivity in the hexagonal perovskite derivative Ba3MoNbO8.5. J Am Chem Soc 138(51):16764–16769
24. Fujii K, Shiraiwa M, Esaki Y, Yashima M, Kim SJ, Lee S (2015) Improved oxide-ion conductivity of NdBaInO4 by Sr doping. J Mater Chem A 3(22):11985–11990
25. Yang X, Liu S, Lu F, Xu J, Kuang X (2016) Acceptor doping and oxygen vacancy migration in layered perovskite NdBaInO4-Based mixed conductors. J Phys Chem C 120(12):6416–6426
26. Shiraiwa M, Fujii K, Esaki Y, Kim SJ, Lee S, Yashima M (2017) Crystal structure and oxide-ion conductivity of Ba 1+ x Nd 1–x InO 4–x /2. J Electrochem Soc 164(13):F1392–F1399
27. Dion M, Ganne M, Tournoux M (1981) Nouvelles familles de phases MIMII2Nb3O10 a feuillets 'perovskites.' Mater Res Bull 16(11):1429–1435
28. Jacobson AJ, Johnson JW, Lewandowski JT (1985) Interlayer chemistry between thick transition-metal oxide layers: synthesis and intercalation reactions of K[Ca2Nan-3NbnO3+1] (3 $\leq$ n $\leq$ 7). Inorg Chem 24(23):3727–3729
29. Mccabe EE et al (2015) Proper ferroelectricity in the dion-jacobson material CsBi2Ti2NbO10: experiment and theory. Chem Mater 27(24):8298–8309
30. Il Kim Y, Kim SY (2015) Ionic conductivity of Dion–Jacobson type oxide LiLaTa2O7and oxynitride LiLaTa2O6.15N0.57measured by impedance spectroscopy. Ceram Int 41(2):3318–3323
31. Thangadurai V, Weppner W (2002) ChemInform abstract: determination of the sodium ion transference number of the Dion–Jacobson-Type Layered Perovskite NaCa2Nb3O10 Using ac impedance and dc methods. ChemInform 33(24):1136–1143
32. Hong Y-S, Kim S-J, Kim S-J, Choy J-H (2000) B-site cation arrangement and crystal structure of layered perovskite compounds CsLn2Ti2NbO10 (Ln = La, Pr, Nd, Sm) and CsCaLaTiNb2O10. J Mater Chem 10(5):1209–1214
33. Torii S et al (2011) Super high resolution powder diffractometer at J-PARC. J Phys Soc Japan 80(SUPPL. B):3–6
34. Oishi R et al (2009) Rietveld analysis software for J-PARC. Nucl Instruments Methods Phys Res Sect A Accel Spectrometers Detect Assoc Equip 600(1):94–96
35. Momma K, Izumi F (2011) VESTA 3 for three-dimensional visualization of crystal, volumetric and morphology data. J Appl Crystallogr 44(6):1272–1276
36. Chen H, Wong LL, Adams S (2019) SoftBV – a software tool for screening the materials genome of inorganic fast ion conductors. Acta Crystallogr Sect B Struct Sci Cryst Eng Mater 75(1):18–33
37. Wong LL, Chen HM, Adams S (2015) Sodium-ion diffusion mechanisms in the low cost high voltage cathode material Na2+δFe2-δ/2(SO4)3. Phys Chem Chem Phys 17(14):9186–9193

38. Chen H, Adams S (2017) Bond softness sensitive bond-valence parameters for crystal structure plausibility tests. IUCrJ 4:614–625
39. Brown ID, Altermatt D (1985) Bond-valence parameters obtained from a systematic analysis of the inorganic crystal structure database. Database 244(2):244–247
40. Brese NE, O'Keeffe M (1991) Bond-valence parameters for solids. Acta Crystallogr Sect B 47(2):192–197

Chapter 4
Conclusion and Future Work

Oxide-ion conductors exhibit many applications in various electrochemical devices such as solid-oxide fuel cells, gas sensors, and separation membranes. And the oxide-ion conductivity of a material is strongly dependent on its crystal structure. Therefore, it is important to discover new structural families of oxide-ion conductors.

The research described in this doctoral thesis was focused on the characterization of the electrical and structural properties of the new structural family of oxide-ion conductors belonging to the Dion–Jacobson-type layered perovskite family. In particular, the results have demonstrated that the Dion–Jacobson-type layered perovskite $CsBi_2Ti_2NbO_{10-\delta}$ is an excellent oxide-ion conductor. This is the first report of a fast oxide-ion conductor in Dion–Jacobson-type layered perovskite family. Concentration cell measurements have evidenced an oxide-ion transport number of 1.00 − 0.98 between 873 and 1173 K in air/O_2, 0.97 − 0.95 between 873 and 1173 K in air/N_2, and 0.87 at 873 K in air/5% H_2 in N_2, thus confirming that the conduction in $CsBi_2Ti_2NbO_{10-\delta}$ is oxide ions. Surprisingly, it was found that the bulk conductivities σ_b of Dion–Jacobson phase $CsBi_2Ti_2NbO_{10-\delta}$ were as high as 8.9×10^{-2} S cm^{-1} at 1073 K and 1.5×10^{-2} S cm^{-1} at 873 K, which were higher than those of the conventional yttria-stabilized zirconia (YSZ) and comparable to those of the best oxide-ion conductors.

The structural origin of the high oxide-ion conductivity using the crystal structure of $CsBi_2Ti_2NbO_{10-\delta}$ was investigated. This structure was obtained by the Rietveld refinement of the neutron-diffraction data measured in situ with SuperHRPD. The refined equivalent isotropic atomic displacement parameters of equatorial oxygen O1 (U_{eq}(O1)) and apical oxygen O2 (U_{eq}(O2)) in the oxide-ion conducting inner perovskite layer were much higher than those of equatorial O3 and apical O4 atoms in the outer perovskite layers. Therefore, the O1 and O2 atoms in the inner oxide-ion conducting perovskite layer exhibited larger thermal motions than the O3 and O4 atoms in the outer perovskite layer. In particular, the anisotropic atomic displacement parameters (U_{ij}) of O1 and O2 atoms were extremely large. The large U_{33}(O1)

W. Zhang, *Oxide-Ion Conduction in the Dion–Jacobson-Type Layered Perovskites*, Springer Theses, https://doi.org/10.1007/978-981-19-2247-3_4

and U_{11}(O2) indicate high anisotropic thermal motions along the *c*- and *a*-axes, respectively, which suggests the O1–O2 oxide-ion diffusion. Similarly, the large U_{11}(O1) means high thermal motions along the *a*- and *b*-axes, indicating the O1–O1 oxide-ion diffusion. The large thermal parameters of the O1 and O2 are attributable to the large bottlenecks in the inner perovskite layer by the large size of Cs^+ and Bi^{3+} displacement. Isosurfaces of neutron scattering length density obtained by the maximum-entropy method (MEM) and bond valence-based energy landscapes also showed anisotropic thermal motions of O1 and O2 atoms. In addition, the O2–O3 diffusion path was observed in the BVELs. We attribute the high oxide-ion conductivity of $CsBi_2Ti_2NbO_{10-\delta}$ not only to the existence of oxygen vacancies but also to the 2D oxide-ion diffusion and the extremely high U_{eq}(O1), U_{eq}(O2), U_{11}(O1), U_{33}(O1), and U_{11}(O2) [= U_{22}(O2)].

A new concept was also proposed; large bottlenecks for oxide-ion migration by the large size of Cs^+ and Bi^{3+} displacement. The Cs^+ expands the Bi–Bi distance along the *b*-axis due to the cation ordering of larger $Cs^+$and smaller Bi^{3+}. The Bi^{3+} cation is displaced along the *c*-axis apart from the Ti/Nb1–O1 layer, which can be explained by the electrostatic forces. The Bi^{3+} displacement expands the Bi–Bi distance along the *c*-axis. The expansions of Bi–Bi distances along both the b (= a) and c axes lead to large bottleneck sizes for O1–O2 (yellow areas) and O1–O1 (pink areas) oxide-ion diffusion.

In order to improve the practice application of the Dion–Jacobson-type layered perovskite $CsBi_2Ti_2NbO_{10-\delta}$, the $CsLa_2Ti_2NbO_{10}$ was investigated which showed no phase transition between room temperature and 1073 K. The oxide-ion conduction was also confirmed in the composition. Unfortunately, because of the low level of oxygen vacancies concentration, the oxide-ion conductivity of $CsLa_2Ti_2NbO_{10}$ was lower than that of $CsBi_2Ti_2NbO_{10-\delta}$.

There are also many future works to be done on the basis of the results presented in this doctoral thesis. The predominance of oxide ion conduction in $CsBi_2Ti_2NbO_{10-\delta}$ could be further confirmed by ^{18}O tracer diffusion coefficient measurements, which have been planned in Prof. S. Skinner's group at Imperial College, London. Furthermore, the $P(O)_2$ dependence measurements could be repeated in order to obtain additional data points in the intermediate $P(O)_2$ region such as using mixed CO or CO_2 gases. Neutron diffraction should be performed under pure O_2, air, N_2, and H2 atmospheres to study the slight structural changes in different atmospheres. Various strategies could be attempted to improve the conductivity and the stability of $CsBi_2Ti_2NbO_{10-\delta}$. At last, aliovalent and substitutions should be investigated as well like the investigation of $CsBi_{1-x}La_xTi_2NbO_{10-\delta}$ compounds, which are expected to own both of the high oxide-ion conductivity from $CsBi_2Ti_2NbO_{10-\delta}$ and the high phase stability without phase transition from $CsLa_2Ti_2NbO_{10}$.

In conclusion, the results presented in this thesis have demonstrated that $CsBi_2Ti_2NbO_{10-\delta}$ is the first example of Dion–Jacobson type layered perovskite to exhibit high oxide-ion conductivity. The bulk conductivity of $CsBi_2Ti_2NbO_{10-\delta}$ (δ is the oxygen vacancy content) was 8.9×10^{-2} S cm^{-1} at 1073 K, which was higher than that of the conventional yttria-stabilized zirconia (YSZ). The structural characteristics at the basis of the oxide ion conduction have been determined and components of oxide-ion conductivity have also been individuated in other Dion–Jacobson-type layered perovskite exhibiting similar structural features. The present finding of high oxide-ion conductivities in the new structure family, Dion–Jacobson-type $CsBi_2Ti_2NbO_{10-\delta}$, and the concept would facilitate the design of novel oxide-ion conductors based on the Dion–Jacobson phases.